T0254200

SpringerBriefs in Physics

More information about this series at http://www.springer.com/series/8902

Hai-Peng Li · Rui-Qin Zhang

Phonon Thermal Transport in Silicon-Based Nanomaterials

 Springer

Hai-Peng Li
School of Physical Science
   and Technology
China University of Mining
   and Technology
Xuzhou, China

Rui-Qin Zhang
Department of Physics
City University of Hong Kong
Hong Kong, China

ISSN 2191-5423   ISSN 2191-5431   (electronic)
SpringerBriefs in Physics
ISBN 978-981-13-2636-3   ISBN 978-981-13-2637-0   (eBook)
https://doi.org/10.1007/978-981-13-2637-0

Library of Congress Control Number: 2018955167

This Springer imprint is published by the registered company Springer Nature Singapore Pte Ltd.
The registered company address is: 152 Beach Road, #21-01/04 Gateway East, Singapore 189721,
Singapore

Foreword

Silicon is a semiconductor material that is widely used in electronics and photovoltaics. In particular, low-dimensional silicon-based nanomaterials are being increasingly adopted in various silicon-based nanotechnologies, thus attracting great attention to the thermal transport properties of these materials due to their nanoelectronic and thermoelectric applications. In silicon-based nanomaterials, thermal energy is predominantly transported by phonons (i.e., quantized lattice vibrations). A greater understanding of phonon transport in silicon-based nanomaterials is strategically important to achieving the optimal use of silicon-based nanostructures in various thermal-related applications. In the past decade, Dr. Hai-Peng Li and Prof. Rui-Qin Zhang have systematically investigated the thermal properties and phonon transport in silicon-based nanomaterials, including nanoclusters, nanowires, and nanosheets, through the use of computational methods. Their research findings on the thermal transport properties of silicon-based nanomaterials are summarized in this SpringerBriefs entitled "Phonon Thermal Transport in Silicon-Based Nanomaterials".

In this book, the authors, Hai-Peng Li and Rui-Qin Zhang, first provide both theorists and experimentalists with an introduction to the current computational and experimental techniques of nanoscale thermal transport and then cover the applications of silicon-based nanostructures in renewable energy, such as thermoelectrics. The book is organized into six chapters. Chapter 1 introduces the research background, briefly reviews the recent research advances, and presents the objectives of the book. Chapter 2 presents a review of the theoretical and experimental methods for determining the thermal conductivity of nanostructures available in the literature. Chapters 3, 4, and 5 provide comprehensive reviews of the theoretical studies on this particular topic, which were partially performed by the authors over recent years. These studies cover the following specific subject matters: thermal stability and phonon thermal transport in spherical silicon nanoclusters; phonon thermal transport in silicon nanowires and its surface effects; and phonon thermal transport in silicene and effects due to defects therein. Chapter 6 presents the summary and concluding remarks.

This book offers a timely and extensive introduction to the current and commonly used theoretical and experimental techniques for determining the thermal conductivity of nanomaterials. The authors' research findings presented in this book also highlight the immense potential of various silicon-based nanomaterials for applications in electronics, thermoelectrics, and solar thermal technologies. With content relevant to both academic and practical points of view, the book will interest researchers and postgraduates as well as technologists in the nanoscience and renewable energy areas. Therefore, the book should be a highly valuable reference for the community of researchers in the fields of condensed matter physics and nanomaterials science and for postgraduate students who would like to learn these techniques and applications.

Hong Kong, China   Michel A. Van Hove
July 2018   Fellow of the APS, AVS, and IoP
                                                                            Head and Chair Professor of the
                                                                                       Department of Physics
                                                                            Hong Kong Baptist University

Acknowledgements

The preparation for the writing of this book can be traced back to the time of my education, and there is a long list of people I wish to thank. I am grateful to my Ph.D. supervisor, Prof. Rui-Qin Zhang from the City University of Hong Kong, for his collaborative research and support in the past years. I must thank him for his help in the preparation of the book proposal and also for coediting the book manuscript. At the time the book manuscript was completed, I was a visiting scholar at the University of Colorado Boulder, USA, financially supported by the China Scholarship Council (Grant No. 201706425053). In addition, I also want to thank my wife and children and my family for their concern and encouragement over the past decades.

Last but not least, I am most grateful to Dr. Jian Li, Senior Editor of Physics and Astronomy at Springer Nature Beijing office, China, for his help in getting the project started. I also wish to thank Morgane Ma from Springer Nature Beijing office, China, for her kind and efficient assistance in editing this book. Finally, I gratefully acknowledge the financial support I received from the National Natural Science Foundation of China (Grant Nos. 11504418, 11447033, 11347123) and the Fundamental Research Funds for the Central Universities of China (Grant No. 2015XKMS075).

Summer 2018   Hai-Peng Li

Contents

Chapter 1
Introduction

The past decades have witnessed rapid progress in the silicon-based semiconductor industry driven by the constant demand for an improvement in the performance and functionality of silicon integrated circuits (ICs) and lower manufacturing costs. This progress is attributed to the significant advances in micro/nanofabrication technology that have resulted in the continuous miniaturization of microelectronic devices. Since the 1960s, the feature size of semiconductor transistors has been shrinking 30% every three years, and the number of transistors in an IC doubles roughly every two years, as governed by Moore's law [1]. By the end of 2017, Intel Corporation began producing a 10 nm chip that can pack 100 million transistors per square millimeter. The present 10 and 7 nm technologies are not the end. Some leading manufactures keep discovering newer technologies to uphold Moore's law to its maximum possibility. So, the trend toward miniaturization will continue in microelectronic chips and will also apply to other technologies, such as cell phones, larger memory banks, digital products, and so on.

The temperature at the chip level within silicon microprocessors rapidly increases because of the increased transistor package density and the shrinking device dimensions in electronic circuits [2]. Particularly, the rapidly increasing on-die heat generation can result in on-die hot spots which adversely affect the performance and reliability of the device. As such, heat dissipation and heat management have become the "bottleneck" in the development of computer chips. Similar thermal issues have also been encountered in optoelectronic and photonic devices, such as infrared detectors and semiconductor lasers, the device material properties of which are often temperature sensitive. Therefore, the search for materials that conduct heat well has become very essential for the design of the next generation of ICs and nanoelectronic devices. Addressing the thermal issues in nanoelectronic devices requires better understanding and controlling of the energy dissipation and thermal transport in nanoscale devices, and this need has recently attracted the attention of scientists all over the world.

© The Author(s), under exclusive licence to Springer Nature Singapore Pte Ltd. 2018
H.-P. Li and R.-Q. Zhang, *Phonon Thermal Transport in Silicon-Based Nanomaterials*,
SpringerBriefs in Physics, https://doi.org/10.1007/978-981-13-2637-0_1

   This renewed interest in nanoscale heat transport also has important implications for many other technologies, an important example being thermoelectric (TE) cooling and power generation based on the TE effect [3]. The TE effect refers to the direct conversion of temperature differences to electric voltages and vice versa (Fig. 1.1). TE generators can convert the heat generated by various sources, including solar radiation, automotive exhaust, and industrial processes, into electricity. A TE refrigerator has no moving parts and does not require a Freon refrigerant. Pop et al. [4] developed a novel strategy for keeping computer chips cool using a chip-scale TE cooler and reported that a TE cooler can lower the temperature of a small hot spot on a large chip by nearly 15 °C. TE devices have many advantages, such as compact size, energy-saving properties, and environmental protection, and are thus attracting considerable attention in the research community that aims to solve current energy challenges.

   The efficiency of TE devices is characterized by the figure of merit, $ZT = \alpha^2 \sigma T / \kappa$, of the material, where α is the Seebeck coefficient, σ is the electrical conductivity, T is the absolute temperature, and κ is the total thermal conductivity. The larger the value of ZT, the better the efficiency of the TE cooler or power generator. Devices with ZT > 1 are considered good TEs; however, a ZT > 3 is required for commercial use. While there is no theoretical limit for the maximum ZT, the best bulk TE materials found so far, such as Bi_2Te_3, PbTe, and $Si_{1-x}Ge_x$, have shown a maximum ZT value of about 1 (Fig. 1.2), which restricts the large-scale application of TE technology [5]. Therefore, finding an effective way to improve the TE efficiency has become a key issue in current TE research.

   However, the high TE conversion efficiency, governed by the dimensionless figure of merit ZT, requires simultaneously high electrical conductivity and Seebeck coefficient yet low thermal conductivity, which are rather difficult to obtain in a single-phase material due to the fact that these physical property parameters are all inter-related [6]. Figure 1.3 provides a guide for the evolution of the different parameters versus the carrier concentration [3]. For example, it is practically impossible to independently modulate both the electrical conductivity and the Seebeck coefficient. They also evolve in the opposite direction when changing the carrier concentration. On the other hand, the thermal conductivity changes very little over small concentration range and increases remarkably above a certain value. Therefore, identifying materials with a high thermoelectric ZT has proven to be an extremely challenging task. After 30 years of slow progress, TE materials research experienced a resurgence in the early 21st century, as shown in Fig. 1.2, inspired by the development of new concepts and theories to engineer electron and phonon transport in both bulk and nanostructured materials [7]. For example, the "phonon-glass, electron-crystal" concept remains a strong driver in TE materials research, resulting in new high-performance materials and an increased focus on controlling structure and chemical bonding to minimize irreversible heat transport in crystals [8, 9]. In nanostructures, quantum and classical size effects provide opportunities to tailor the electron and phonon transport through structural engineering. Quantum wells, superlattices, quantum wires, and quantum dots have been utilized to tune the band structure, energy levels, and density of states of electrons and have led to the improved energy conversion

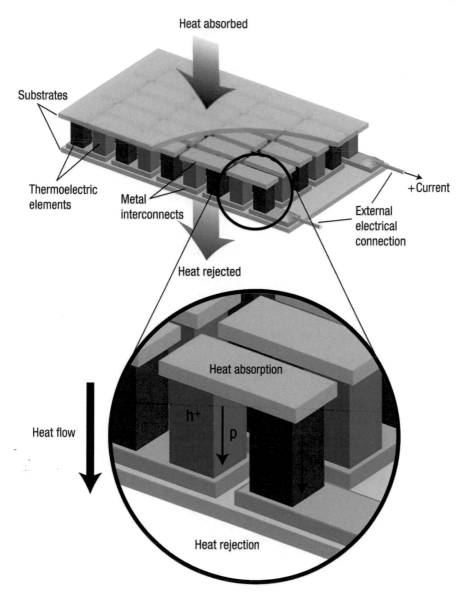

Fig. 1.1 Schematic of TE module showing the direction of charge flow in cooling and power generation. Reprinted with permission from Ref. [3], copyright (2008) by Springer Nature

capability of charged carriers compared to that of their various bulk counterparts [10–12]. Interface reflection and the scattering of phonons in these nanostructures have been employed to reduce the thermal conductivity [13–15]. Increases in the TE figure of merit based on size effects for either electrons or phonons have been demonstrated [16, 17].

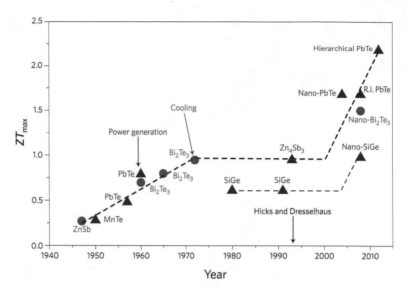

Fig. 1.2 Evolution of the maximum ZT over time for popular or prospective TE materials. Materials for TE cooling are shown as blue dots, and materials for TE power generation are shown as red triangles. Reprinted with permission from Ref. [5], copyright (2013) by Springer Nature

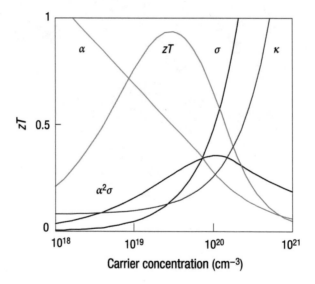

Fig. 1.3 Evolution of the Seebeck coefficient α (blue), the electrical conductivity σ (brown), and the thermal conductivity κ (violet), as well as the resulting power factor (black) and figure of merit (green) versus carrier concentration. The trends are modelled for Bi_2Te_3. Reprinted with permission from Ref. [3], copyright (2008) by Springer Nature

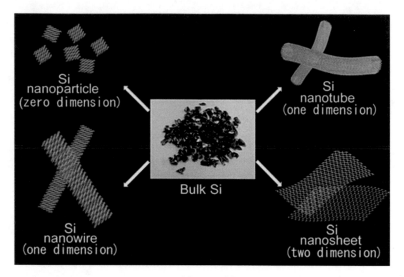

Fig. 1.4 Schematic illustration of low-dimensional silicon nanomaterials. Reprinted with permission from Ref. [22], copyright (2011) by John Wiley & Sons

Contrary to thermoelectricity, which requires the lowest possible heat conductivity, heat dissipation in electronic chips requires the opposite, that is, a higher heat dissipation rate [18]. In order to achieve any of these contradictory effects, a better understanding of thermal transport mechanisms is required. Lastly, phonon transport represents just one aspect of a larger understanding about phonons which also includes their generation, detection, and manipulation [19]. While this field is lagging behind its photonic counterpart, a strong research effort in this direction offers interesting prospects [20].

The development of nanotechnology has enabled the fabrication of materials with structures that vary on the order of a few nanometers by using the top-down approach as well as the bottom-up strategy [21, and references therein]. For instance, to date, as shown in Fig. 1.4, various silicon low-dimensional nanostructures [22], such as zero-dimensional (0D) silicon nanoparticles [23, 24], one-dimensional (1D) silicon nanotubes [25, 26] and nanowires [27, 28], and two-dimensional (2D) silicon nanosheets [29–31], have been synthesized. Thanks to the advances in nanofabrication, it has become easier to study the impact of such small dimensions on the thermal transport in silicon-based nanomaterials. This direction of research has produced encouraging results, and a number of publications on theoretical and experimental studies—on superlattices and nanowires initially—have emerged [32–40].

In 2003, Li et al. [32] experimentally reported that the room-temperature thermal conductivity of single crystalline silicon nanowires is one or two orders of magnitude lower than that of bulk silicon. They also observed a strong diameter dependence. Surface roughness has also been experimentally shown to further reduce the thermal conductivity of nanowires, resulting in a 100-fold improvement in the TE efficiency compared with that of bulk silicon at room temperature [16]. Such experimental findings have resulted in great interest in research on phonon thermal transport in nano-silicon. Extensive studies on the thermal conductivity of various silicon nanowires, such as core-shell nanowires [33], surface-decorated nanowires [34], and superlatticed nanowires [35, 36], have now been conducted using theoretical methods as well as experimental measurements.

Heat transport in thin films has also received considerable attention. Experimental and theoretical studies [37–42] have been conducted to investigate the thermal conductivity of the various silicon nanofilms used in microelectronics, photonics, and TEs. Our research team and others have been theoretically studying the thermal conductivity of silicene, which is a graphene-like monolayer structure of silicon. A few strategies have also been used to reduce thermal transport in silicon nanosheets, including adding substrates [43, 44], heterostructuring [45], surface functionalization [46], and defects [47–51]. These studies provide significant guidance for experimental realization.

To date, studies on the thermal transport in silicon nanoparticles are fewer than those on nanowires and nanosheets; however, these studies will become increasingly important because of the developments in nanoparticle applications [52, 53]. For example, SiGe nanocomposites embedded in the germanium (Ge) matrix are promising candidates for TE devices [10, 54]. More recently, periodic patterning has attracted much attention in relation to controlling heat flow with phononic crystals in similar ways to how light is controlled with photonic crystals [55, 56].

Over the past decades, we and coworkers have performed systematic theoretical studies on different aspects of silicon-based nanomaterials, such as structures, growth mechanisms, and properties, aiming to fill the deficiencies in the field. In the SpringerBriefs [57] published in 2014, Prof. R. Q. Zhang provided a comprehensive review of his group's quantum-mechanical calculations on various possible silicon nanostructures for their growth mechanism, structural and chemical stability, excited-state property, and energy-band engineering. Following this previous book, in the present Brief, we focus on reviewing our recent molecular dynamics studies on the thermal properties of silicon-based nanomaterials [18, 48, 58–62] such as 0D silicon clusters, 1D silicon nanowires, and 2D silicene. The book specifically includes the size effect of thermal stability and phonon thermal transport in spherical silicon nanoclusters, the surface effects of phonon thermal transport in silicon nanowires, and the defects effects of phonon thermal transport in silicene. The findings presented in this book can provide insight into and predict the thermal transport properties in silicon-based nanomaterials critically needed by experimentalists. Additionally, we specifically review the advance in theoretical and experimental methods for determining the thermal conductivity of nanostructures available in the literature; this review will provide a very helpful reference for young researchers, including graduate-level students getting ready to embark on research in this new field.

References

1. Pease, R.F., Chou, S.Y.: Lithography and other patterning techniques for future electronics. Proc. IEEE **96**(2), 248–270 (2008). https://doi.org/10.1109/JPROC.2007.911853
2. Mahajan, R., Chiu, C.P., Chrysler, G.: Cooling a microprocessor chip. Proc. IEEE **94**(8), 1476–1486 (2006). https://doi.org/10.1109/JPROC.2006.879800
3. Snyder, G.J., Toberer, E.S.: Complex thermoelectric materials. Nat. Mater. **7**(2), 105 (2008). https://doi.org/10.1038/nmat2090
4. Pop, E., Sinha, S., Goodson, K.E.: Heat generation and transport in nanometer-scale transistors. Proc. IEEE **94**(8), 1587–1601 (2006). https://doi.org/10.1109/JPROC.2006.879794
5. Heremans, J.P., Dresselhaus, M.S., Bell, L.E., Morelli, D.T.: When thermoelectrics reached the nanoscale. Nat. Nanotechnol. **8**, 471–473 (2013). https://doi.org/10.1038/nnano.2013.129
6. Esfahani, E.N., Ma, F., Wang, S., Ou, Y., Yang, J., Li, J.: Quantitative nanoscale mapping of three-phase thermal conductivities in filled skutterudites via scanning thermal microscopy. Nat. Sci. Rev. **5**(1), 59–69 (2018). https://doi.org/10.1093/nsr/nwx074
7. Chen, G., Dresselhaus, M.S., Dresselhaus, G., Fleurial, J.P., Caillat, T.: Recent developments in thermoelectric materials. Int. Mater. Rev. **48**(1), 45–66 (2003). https://doi.org/10.1179/095 066003225010182
8. Takabatake, T., Suekuni, K., Nakayama, T., Kaneshita, E.: Phonon-glass electron-crystal thermoelectric clathrates: experiments and theory. Rev. Mod. Phys. **86**(2), 669 (2014). https://doi.org/10.1103/RevModPhys.86.669
9. Beekman, M., Morelli, D.T., Nolas, G.S.: Better thermoelectrics through glass-like crystals. Nat. Mater. **14**(12), 1182 (2015). https://doi.org/10.1038/nmat4461
10. Dresselhaus, M.S., Chen, G., Tang, M.Y., Yang, R.G., Lee, H., Wang, D.Z., Ren, Z.F., Fleurial, J.P., Gogna, P.: New directions for low-dimensional thermoelectric materials. Adv. Mater. **19**(8), 1043–1053 (2007). https://doi.org/10.1002/adma.200600527
11. Bulusu, A., Walker, D.G.: Modeling of thermoelectric properties of semiconductor thin films with quantum and scattering effects. J. Heat Transf. **129**(4), 492–499 (2007). https://doi.org/1 0.1115/1.2709962
12. Dresselhaus, M.S., Dresselhaus, G., Sun, X., Zhang, Z., Cronin, S.B., Koga, T.: Low-dimensional thermoelectric materials. Phys. Solid State **41**(5), 679–682 (1999). https://doi.org/10.1134/1.1130849
13. Zhou, Y., Yao, Y., Hu, M.: Boundary scattering effect on the thermal conductivity of nanowires. Semicond. Sci. Tech. **31**(7), 074004 (2016). https://doi.org/10.1088/0268-1242/31/7/074004
14. Zhou, G., Zhang, G.: General theories and features of interfacial thermal transport. Chin. Phys. B **27**(3), 034401 (2018). https://doi.org/10.1088/1674-1056/27/3/034401
15. Xu, W., Zhang, G., Li, B.: Interfacial thermal resistance and thermal rectification between suspended and encased single layer graphene. J. Appl. Phys. **116**(13), 134303 (2014). https://doi.org/10.1063/1.4896733
16. Hochbaum, A.I., Chen, R., Delgado, R.D., Liang, W., Garnett, E.C., Najarian, M., Majumdar, A., Yang, P.: Enhanced thermoelectric performance of rough silicon nanowires. Nature **451**(7175), 163 (2008). https://doi.org/10.1038/nature06381
17. Boukai, A.I., Bunimovich, Y., Tahir-Kheli, J., Yu, J.K., Goddard III, W.A., Heath, J.R.: Silicon nanowires as efficient thermoelectric materials. Nature **451**(7175), 168 (2008). https://doi.org/10.1038/nature06458
18. Li, H.P.: Molecular dynamics simulations of phonon thermal transport in low-dimensional silicon structures. Doctoral dissertation, City University of Hong Kong (2012)
19. Li, N., Ren, J., Wang, L., Zhang, G., Hänggi, P., Li, B.: Phononics: manipulating heat flow with electronic analogs and beyond. Rev. Mod. Phys. **84**(3), 1045 (2012). https://doi.org/10.1 103/RevModPhys.84.1045
20. Xu, X., Zhou, J., Yang, N., Li, N., Li, Y., Li, B.: Artificial microstructure materials and heat flux manipulation. Sci. Sinica Technol. **45**(7), 705 (2015). https://doi.org/10.1360/N092015-0 0122

21. Teo, B.K., Sun, X.H.: Silicon-based low-dimensional nanomaterials and nanodevices. Chem. Rev. **107**(5), 1454–1532 (2007). https://doi.org/10.1021/cr030187n
22. Okamoto, H., Sugiyama, Y., Nakano, H.: Synthesis and modification of silicon nanosheets and other silicon nanomaterials. Chemistry **17**(36), 9864–9887 (2011). https://doi.org/10.1002/chem.201100641
23. Bley, R.A., Kauzlarich, S.M.: A low-temperature solution phase route for the synthesis of silicon nanoclusters. J. Am. Chem. Soc. **118**(49), 12461–12462 (1996). https://doi.org/10.1021/ja962787s
24. Van Buuren, T., Dinh, L.N., Chase, L.L., Siekhaus, W.J., Terminello, L.J.: Changes in the electronic properties of Si nanocrystals as a function of particle size. Phys. Rev. Lett. **80**(17), 3803 (1998). https://doi.org/10.1103/PhysRevLett.80.3803
25. De Crescenzi, M., Castrucci, P., Scarselli, M., Diociaiuti, M., Chaudhari, P.S., Balasubramanian, C., Bhave, T.M., Bhoraskar, S.V.: Experimental imaging of silicon nanotubes. Appl. Phys. Lett. **86**(23), 231901 (2005). https://doi.org/10.1063/1.1943497
26. Perepichka, D.F., Rosei, F.: Silicon nanotubes. Small **2**(1), 22–25 (2006). https://doi.org/10.1002/smll.200500276
27. Zhang, R.Q., Lifshitz, Y., Lee, S.T.: Oxide-assisted growth of semiconducting nanowires. Adv. Mater. **15**(7–8), 635–640 (2003). https://doi.org/10.1002/adma.200301641
28. Morales, A.M., Lieber, C.M.: A laser ablation method for the synthesis of crystalline semiconductor nanowires. Science **279**(5348), 208–211 (1998). https://doi.org/10.1126/science.279.5348.208
29. Okamoto, H., Kumai, Y., Sugiyama, Y., Mitsuoka, T., Nakanishi, K., Ohta, T., Nozaki, H., Yamaguchi, S., Shirai, S., Nakano, H.: Silicon nanosheets and their self-assembled regular stacking structure. J. Am. Chem. Soc. **132**(8), 2710–2718 (2010). https://doi.org/10.1021/ja908827z
30. Aufray, B., Kara, A., Vizzini, S., Oughaddou, H., Leandri, C., Ealet, B., Le Lay, G.: Graphene-like silicon nanoribbons on Ag (110): a possible formation of silicene. Appl. Phys. Lett. **96**(18), 183102 (2010). https://doi.org/10.1063/1.3419932
31. Zhang, C., De Sarkar, A., Zhang, R.Q.: Strain induced band dispersion engineering in Si nanosheets. J. Phys. Chem. C **115**(48), 23682–23687 (2011). https://doi.org/10.1021/jp206911b
32. Li, D., Wu, Y., Kim, P., Shi, L., Yang, P., Majumdar, A.: Thermal conductivity of individual silicon nanowires. Appl. Phys. Lett. **83**(14), 2934–2936 (2003). https://doi.org/10.1063/1.1616981
33. Sarikurt, S., Ozden, A., Kandemir, A., Sevik, C., Kinaci, A., Haskins, J.B., Cagin, T.: Tailoring thermal conductivity of silicon/germanium nanowires utilizing core-shell architecture. J. Appl. Phys. **119**(15), 155101 (2016). https://doi.org/10.1063/1.4946835
34. Markussen, T., Jauho, A.P., Brandbyge, M.: Surface-decorated silicon nanowires: a route to high-ZT thermoelectrics. Phys. Rev. Lett. **103**(5), 055502 (2009). https://doi.org/10.1103/PhysRevLett.103.055502
35. Li, D., Wu, Y., Fan, R., Yang, P., Majumdar, A.: Thermal conductivity of Si/SiGe superlattice nanowires. Appl. Phys. Lett. **83**(15), 3186–3188 (2003). https://doi.org/10.1063/1.1619221
36. Mu, X., Wang, L., Yang, X., Zhang, P., To, A.C., Luo, T.: Ultra-low thermal conductivity in Si/Ge hierarchical superlattice nanowire. Sci. Rep. **5**, 16697 (2015). https://doi.org/10.1038/srep16697
37. Chen, G.: Particularities of heat conduction in nanostructures. J. Nanopart. Res. **2**(2), 199–204 (2000). https://doi.org/10.1023/A:1010003718481
38. Ju, Y.S., Goodson, K.E.: Phonon scattering in silicon films with thickness of order 100 nm. Appl. Phys. Lett. **74**(20), 3005–3007 (1999). https://doi.org/10.1063/1.123994
39. Wang, Z., Li, Z.: Lattice dynamics analysis of thermal conductivity in silicon nanoscale film. Appl. Therm. Eng. **26**(17–18), 2063–2066 (2006). https://doi.org/10.1016/j.applthermaleng.2006.04.020
40. Terris, D., Joulain, K., Lemonnier, D., Lacroix, D., Chantrenne, P.: Prediction of the thermal conductivity anisotropy of Si nanofilms. Results of several numerical methods. Int. J. Therm. Sci. **48**(8), 1467–1476 (2009). https://doi.org/10.1016/j.ijthermalsci.2009.01.005

41. Liu, W., Asheghi, M.: Phonon-boundary scattering in ultrathin single-crystal silicon layers. Appl. Phys. Lett. **84**(19), 3819–3821 (2004). https://doi.org/10.1063/1.1741039
42. Liu, W., Asheghi, M.: Thermal conductivity measurements of ultra-thin single crystal silicon layers. J. Heat Transfer **128**(1), 75–83 (2006). https://doi.org/10.1115/1.2130403
43. Wang, Z., Feng, T., Ruan, X.: Thermal conductivity and spectral phonon properties of free-standing and supported silicene. J. Appl. Phys. **117**(8), 084317 (2015). https://doi.org/10.1063/1.4913600
44. Zhang, X., Bao, H., Hu, M.: Bilateral substrate effect on the thermal conductivity of two-dimensional silicon. Nanoscale **7**(14), 6014–6022 (2015). https://doi.org/10.1039/C4NR06523A
45. Wang, X., Hong, Y., Chan, P.K., Zhang, J.: Phonon thermal transport in silicene-germanene superlattice: a molecular dynamics study. Nanotechnology **28**(25), 255403 (2017). https://doi.org/10.1088/1361-6528/aa71fa
46. Liu, Z., Wu, X., Luo, T.: The impact of hydrogenation on the thermal transport of silicene. 2D Mater. **4**, 025002 (2017). https://doi.org/10.1088/2053-1583/aa533e
47. Liu, B., Reddy, C.D., Jiang, J., Zhu, H., Baimova, J.A., Dmitriev, S.V., Zhou, K.: Thermal conductivity of silicene nanosheets and the effect of isotopic doping. J. Phys. D Appl. Phys. **47**(16), 165301 (2014). https://doi.org/10.1088/0022-3727/47/16/165301
48. Li, H.P., Zhang, R.Q.: Vacancy-defect-induced diminution of thermal conductivity in silicene. EPL **99**(3), 36001 (2012). https://doi.org/10.1209/0295-5075/99/36001
49. Sadeghi, H., Sangtarash, S., Lambert, C.J.: Enhanced thermoelectric efficiency of porous silicene nanoribbons. Sci. Rep. **5**, 9514 (2015). https://doi.org/10.1038/srep09514
50. Zhao, W., Guo, Z.X., Zhang, Y., Ding, J.W., Zheng, X.J.: Enhanced thermoelectric performance of defected silicene nanoribbons. Solid State Commun. **227**, 1–8 (2016). https://doi.org/10.1016/j.ssc.2015.11.012
51. Wirth, L.J., Osborn, T.H., Farajian, A.A.: Resilience of thermal conductance in defected graphene, silicene, and boron nitride nanoribbons. Appl. Phys. Lett. **109**(17), 173102 (2016). https://doi.org/10.1063/1.4965294
52. Fang, K.C., Weng, C.I., Ju, S.P.: An investigation into the structural features and thermal conductivity of silicon nanoparticles using molecular dynamics simulations. Nanotechnology **17**(15), 3909 (2006). https://doi.org/10.1088/0957-4484/17/15/049
53. Ashby, S.P., Thomas, J.A., García-Cañadas, J., Min, G., Corps, J., Powell, A.V., Xu, H., Shen, W., Chao, Y.: Bridging silicon nanoparticles and thermoelectrics: phenylacetylene functionalization. Faraday Discuss. **176**, 349–361 (2015). https://doi.org/10.1039/C4FD00109E
54. Huang, X., Huai, X., Liang, S., Wang, X.: Thermal transport in Si/Ge nanocomposites. J. Phys. D Appl. Phys. **42**(9), 095416 (2009). https://doi.org/10.1088/0022-3727/42/9/095416
55. Davis, B.L., Hussein, M.I.: Nanophononic metamaterial: thermal conductivity reduction by local resonance. Phys. Rev. Lett. **112**(5), 055505 (2014). https://doi.org/10.1103/PhysRevLett.112.055505
56. Hopkins, P.E., Reinke, C.M., Su, M.F., Olsson III, R.H., Shaner, E.A., Leseman, Z.C., Serrano, J.R., Phinney, L.M., El-Kady, I.: Reduction in the thermal conductivity of single crystalline silicon by phononic crystal patterning. Nano Lett. **11**(1), 107–112 (2010). https://doi.org/10.1021/nl102918q
57. Zhang, R.Q.: Growth Mechanisms and Novel Properties of Silicon Nanostructures from Quantum-Mechanical Calculations. Springer, Berlin Heidelberg (2013). https://doi.org/10.1007/978-3-642-40905-9
58. Li, H.P., De Sarkar, A., Zhang, R.Q.: Surface-nitrogenation-induced thermal conductivity attenuation in silicon nanowires. EPL **96**(5), 56007 (2011). https://doi.org/10.1209/0295-5075/96/56007
59. Li, H.P., Zhang, R.Q.: Size-dependent structural characteristics and phonon thermal transport in silicon nanoclusters. AIP Adv. **3**(8), 082114 (2013). https://doi.org/10.1063/1.4818591
60. Li, H.P., Zhang, R.Q.: Anomalous effect of hydrogenation on phonon thermal conductivity in thin silicon nanowires. EPL **105**(5), 56003 (2014). https://doi.org/10.1209/0295-5075/105/56003

61. Xu, R.F., Han, K., Li, H.P.: Effect of isotope doping on phonon thermal conductivity of silicene nanoribbons: a molecular dynamics study. Chin. Phys. B **27**(2), 026801 (2018). https://doi.org/10.1088/1674-1056/27/2/026801

62. Li, H.P., Zhang, R.Q.: Surface effects on the thermal conductivity of silicon nanowires. Chin. Phys. B **27**(3), 036801 (2018). https://doi.org/10.1088/1674-1056/27/3/036801

Chapter 2
Theoretical and Experimental Methods for Determining the Thermal Conductivity of Nanostructures

2.1 Concepts and Foundations

2.1.1 Thermal Conductivity

Heat transfer by conduction (or heat conduction) refers to the transfer of thermal energy within a material, without any movement of the material as a whole. This transfer occurs when a temperature gradient exists in a solid material. If different parts of an isolated solid are at different temperatures, then thermal energy transfers from the hotter regions to the cooler ones until eventually all areas reach the same temperature. Thermal conductivity is the intrinsic property of a material that describes its ability to conduct heat. That is, heat transfer across materials with high thermal conductivity occurs at a higher rate than that across materials with low thermal conductivity.

The empirical relationship between the conduction rate of a material and the temperature gradient in the direction of energy flow was first formulated in 1822 by Joseph Fourier. Fourier concluded that "*the heat flux resulting from thermal conduction is proportional to the magnitude of the temperature gradient and opposite to it in sign*" [1]. This observation called Fourier's law may be expressed as

$$J = -\kappa \nabla T, \tag{2.1}$$

which resolves itself into three components: $J_x = -\kappa \frac{\partial T}{\partial x}$, $J_y = -\kappa \frac{\partial T}{\partial y}$, and $J_z = -\kappa \frac{\partial T}{\partial z}$. J is the heat flux defined as the amount of heat Q transported through the unit cross-sectional area per unit time (in W/m^2), κ is the thermal conductivity tensor (in W/mK), and ∇T is the temperature gradient (in K/m). The negative sign in Eq. (2.1) indicates that the positive heat transfer direction is opposite the temperature

© The Author(s), under exclusive licence to Springer Nature Singapore Pte Ltd. 2018
H.-P. Li and R.-Q. Zhang, *Phonon Thermal Transport in Silicon-Based Nanomaterials*,
SpringerBriefs in Physics, https://doi.org/10.1007/978-981-13-2637-0_2

gradient. Generally, κ is treated as a constant for small temperature variations but is temperature-dependent in a wide temperature range. κ varies with the crystal orientation and is represented by the tensor $\kappa_{\mu\nu}$ in anisotropic materials. However, κ is reduced to a scalar variable in a material with a cubic isotropy. Fourier's law thus provides the quantitative definition of thermal conductivity for bulk materials and nanostructures [2] and forms the basis of experimental and theoretical methods for determining the thermal conductivity.

2.1.2 Phonon and Electron Contributions

Atoms in a real crystal are not fixed at rigid sites on a lattice but are vibrating. In a periodic structure, vibrations have a waveform with spatial and temporal parts,

$$u(r, t) = u_0 e^{ik \cdot r} e^{-i\omega t}, \tag{2.2}$$

where u_0 is the amplitude, r is the lattice site, k is the wave vector, and ω is the frequency of the vibration modes. A phonon is a quantum mechanical description of a special type of vibrational motion in which a lattice uniformly oscillates at the same frequency. In classical mechanics, this state is known as the normal mode, which is a wave-like phenomenon that exhibits particle-like properties in the wave-particle duality of quantum mechanics. Often referred to as a quasiparticle, a phonon represents a quantized excitation of the modes of vibrations. Similar to an electron, a phonon also has energy and momentum and thus can conduct heat. Phonons play a major role in many of the physical properties of solids, including the thermal and electrical conductivities of the material.

In a solid, thermal energy is carried both by electrons and phonons. As such, the thermal conductivity κ consists of two components, i.e., $\kappa = \kappa_p + \kappa_e$, where κ_p and κ_e denote the phonon (lattice) thermal conductivity and the electron thermal conductivity, respectively. In metals, κ_e is dominant because of the large concentrations of free carriers. The same mobile electrons that participate in electrical conduction also take part in heat transfer. For example, in pure copper (one of the best metallic heat conductors), $\kappa \approx 400$ W/mK at room temperature, and κ_p is limited to 1–2% of the total value [3]. Measurements of the electrical conductivity (σ) define κ_e via the Wiedemann-Franz law, as follows: $\kappa_e/\sigma = LT$, where the constant of proportionality L is called the Lorenz number. By contrast, κ_e in dielectric and undoped semiconductors is negligible, and phonons dominate the heat transfer at low temperatures because of the lack of valence electrons. For example, the electron contribution to the thermal conductivity of bulk silicon has been experimentally determined as approximately 0.65 W/mK at 1000 K (approximately 2% of the total value of 31.0 W/mK) and can be negligible below 1000 K [4, 5]. The prediction of thermal conductivity in nonmetals such as silicon requires the accurate description of phonon behavior or the atomic trajectories in the lattice.

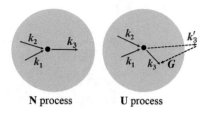

Fig. 2.1 Three-phonon **N** process and **U** process. The grey area denotes the first Brillouin zone. Readapted with permission from Ref. [9], copyright (2015) by Elsevier

N process   U process

2.1.3 Diffusive and Ballistic Phonon Transport

According to quantum mechanics principles, energy carriers in solids, such as phonons and electrons, have both wave-like and particle-like characteristics [6]. Parameters related to the wave-like characteristics of phonons include wavelength and coherence. When the size of the sample is comparable with the phonon wavelength, the standing wave and phonon coherent phenomenon are observed [7]. In addition, the parameter with the particle-like characteristic is the phonon mean free path, Λ, which is the average distance that a phonon particle travels between collisions. Collisions may be caused through phonon scatterings, which mainly involve two categories: normal phonon-phonon scattering (**N** process) and resistive scattering (**R** process). **R** process primarily entails the three-phonon Umklapp process, phonon-boundary scattering, phonon-imperfection scattering, and phonon-electron scattering. Energy conservation is ensured for both kinds of processes, but the quasi-momentum of phonons is conserved only in the **N** process. Taking the three-phonon scattering as an example (Fig. 2.1), the phonon scattering processes satisfy the conservations of energy and quasimomentum,

$$\hbar\omega_1 + \hbar\omega_2 = \hbar\omega_3, \tag{2.3}$$

$$\hbar\mathbf{k}_1 + \hbar\mathbf{k}_2 = \hbar\mathbf{k}_3 + \mathbf{G}, \tag{2.4}$$

where ω_i ($\mathbf{k}_i$) is the frequency (wavevector) of phonon mode i, and $\mathbf{G}$ is the reciprocal lattice vector that is zero for the **N** process and nonzero for the Umklapp (**U**) process. For two-dimensional (2D) lattices such as graphene, an additional selection rule applies because of the reflection symmetry perpendicular to the 2D plane [8, 9].

Depending on the sample size, the Λ value may be comparable or larger than the object size, l. The thermal transport is *diffusive* if l is much larger than Λ, i.e., if phonons undergo numerous scattering events. Among these collisions, the thermal phonon is *hydrodynamic* if the **N** process dominates over the **R** process. By contrast, in low-dimensional nanostructures, such as nanoparticles, nanowires, and nanoscale thin films, l is comparable to or smaller than Λ. In this case, phonons in defectless structures can propagate without scattering, and the thermal conductivity becomes *ballistic* (similar to the *ballistic* electronic conductivity). Therefore, distinguishing between *diffusive* and *ballistic* phonon-transport regimes is essential.

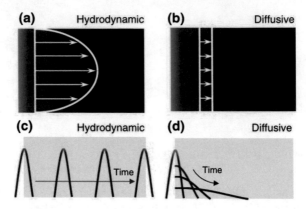

Fig. 2.2 Different macroscopic transport phenomena in *hydrodynamic* and *diffusive* regimes. Reprinted with permission from Ref. [10], copyright (2015) by Springer Nature

Specifically, macroscopic transport behaviors remarkably differ between *hydrodynamic* and *diffusive* regimes [10]. Similar to the mass flux profile of fluid flow in a tube, phonons exhibit a nonuniform heat flux profile in a *hydrodynamic* regime (Fig. 2.2a). The small flux near the boundary is caused by the momentum loss induced by diffuse boundary scattering. However, the heat flux profile of phonons shows a uniform shape in a *diffusive* regime (Fig. 2.2b), because *diffusive* transport resistance is mainly induced by intrinsic momentum loss scattering (i.e., **U** scattering) rather than by diffuse boundary scattering. **U** scattering can exist anywhere inside a sample, thus leading to a uniform heat flux profile. Conversely, in the *hydrodynamic* regime, a heat pulse propagates without significant damping such that the invoked pulse can be transmitted through the sample and finally reach the edge of the sample with an unchanged pulse shape (Fig. 2.2c). In the *diffusive* regime, a heat pulse is severely damped and cannot propagate (Fig. 2.2d), resulting in the pulse reaching the sample edge after a relatively long time.

2.1.4 Kinetic Behavior of Phonon Transport

From kinetic theory, the phonon thermal conductivity of isotropic materials, κ_p, is related to the phonon mean free path (Λ)/relaxation time (τ), as follows:

$$\kappa_p = \frac{1}{3}Cv\tau^2 = \frac{1}{3}Cv\Lambda, \tag{2.5}$$

where C is the average specific heat capacity and v is the average speed of sound. The phonon mean free path is related to the relaxation time with $\Lambda = v\tau$. Although the specific Λ or τ values are difficult to ascertain without some a priori knowledge of thermal transport in a given system, Eq. (2.5) qualitatively explains the fundamen-

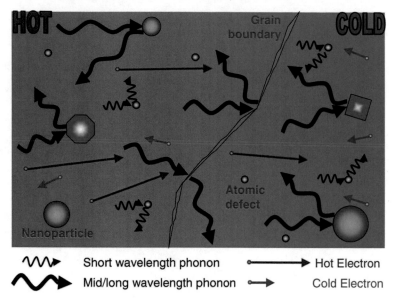

Fig. 2.3 Schematic of illustrating various phonon scattering processes in a nanomaterial. Reprinted with permission from Ref. [13], copyright (2010) by John Wiley & Sons

tal physics and contribution factors of thermal conductivity, which are extremely important in the study of nanoscale thermal transport. A more accurate expression for thermal transport based on quantum mechanics can be found in quantum transport theory [11, 12].

The thermal conductivity in nanostructures is usually less than that in crystals because the existing complex phonon scattering processes limit the mean free path of phonons. As illustrated in Fig. 2.3, atomic defects are effective at scattering short-wavelength phonons, but larger embedded nanoparticles are needed to effectively scatter mid- and long-wavelength phonons. Grain boundaries can also play an effective role in scattering these longer-wavelength phonons. Therefore, from the viewpoint of phonon engineering, reduction of phonon thermal conductivity can be realized using the following major **R** processes: three-phonon Umklapp scattering, mass-impurity scattering, defect (vacancy) scattering, boundary scattering, and phonon-electron scattering. Phonon scattering is a highly frequency-dependent occurrence, thereby making the estimation of the mean free path difficult and prone to uncertainties from complicated systems.

2.2 Theoretical Methods for Determining Nanoscale Thermal Conductivity

2.2.1 Overview of Current Theoretical Methods

Describing thermal conductivity accurately is an important issue for designing nano devices. Treating thermal transport at nanoscale logically entails theoretical and experimental methods. Although several methods have been developed to measure the thermal conductivities of bulk- and nano-structures (as reviewed in the Sect. 2.3), experimental measurements of nanoscale thermal conductivity still face many difficulties [14], such as the existence of contact thermal and contact electric resistances between the sample and substrate, and the complexity of clearly defining temperature under sub-nano sizes. Numerous studies focused on developing theoretical and computational methods. Initially, kinetic modeling based on the phonon Boltzmann transport equation (BTE) and simplified molecular dynamics (MD) simulations were the major tools for identifying phonon thermal conductivity in small model systems. With the advances in computational physics and technology, MD simulations and first-principles calculations have become widely used to predict the thermal transport properties of bulk- and nano-structures in recent years [15]. The non-equilibrium Green's function (NEGF) method is also developed to handle quantum thermal transport at low temperatures. Monte Carlo or lattice Boltzmann method solutions of the phonon BTE supplemented with information from first-principles calculation, lattice dynamics or MD simulations are currently widely used in multi-scale modeling of nanoscale heat transport [16]. The first-principles method or MD approach explains what happens at microscale, whereas kinetic theory-based modeling describes the mesoscopic statistical behavior of phonons.

Many fruitful efforts have also been devoted to the macroscopic description of phonon transport. Based on generalized heat-transport equations, the macroscopic approach can describe thermal transport over a wide range of length scales, including not only the conventional *diffusive* regime (where heat conduction obeys Fourier's law), but also the *ballistic* and the *hydrodynamic* regimes. Actually, four basic macroscopic models for nanoscale heat transport exist in the literature [9, and references therein], namely, the phonon hydrodynamic model, dual-phase-lag model, ballistic-diffusive model, and thermon gas model. Among the available macroscopic methods, the phonon hydrodynamic model is the most well-known model [17]. The denomination "phonon hydrodynamics" classically refers to a particular phonon flow regime with the dominant effect of *normal* phonon-phonon collisions over *resistive* collisions, i.e., the *hydrodynamic* regime. Note that the phonon hydrodynamics is directly produced from the phonon BTE (which can provide a clear and intuitive physical picture); thus, phonon hydrodynamics converts the abstract and ambiguous thermal transport process into a concrete and evident phonon gas flow process [9]. However, a comprehensive account of the theoretical foundations, development, and implementation of this approach in various nanostructures remains lacking [18, 19].

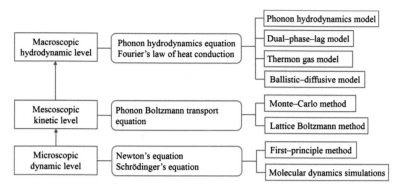

Fig. 2.4 Hierarchical theoretical descriptions of phonon thermal transport at different levels. Drawn according to Ref. [9]

Figure 2.4 summarizes the current theoretical methods for describing nanoscale phonon thermal transport at three different levels of descriptions. The microscopic level provides atomic-scale information. The mesoscopic level produces statistical information such as the particle distribution function. The macroscopic level uses only several state variables for continuum media. Interested readers can refer to review articles and books on the fundamental thermal transport theories of low-dimensional systems [20–26]. Various theoretical approaches have been commonly used to investigate thermal transport in silicon-based nanomaterials, including the phonon BTE [27, 28], MD simulations [29–34], and the NEGF method [35–37].

Next, we briefly review these theoretical approaches.

2.2.2 Phonon Boltzmann Transport Equation

The quasiparticle picture is established for phonons when heat conduction occurs in a dielectric crystal with a characteristic length much larger than the dominant phonon wavelength [19]. Analogous to the transport equation proposed by Boltzmann [38] for rarefied gas transport, the following phonon Boltzmann transport equation is first introduced by Peierls to describe the transport behavior of phonons [39]:

$$\frac{\partial f}{\partial t} + v_g \cdot \nabla f = C(f), \qquad (2.6)$$

where $f = f(r, t, k)$ is the phonon distribution function, with $f(r, t, k)drdk$ denoting the probabilistic number of phonons found within the spatial interval $(r, r + dr)$ and wavevector interval $(k, k + dk)$ at a specific time t. The phonon group velocity v_g signifies the energy propagating speed of the lattice wave and is determined from $v_g = \nabla_k \omega$ as long as the phonon dispersion relation $\omega = \omega(k)$ is available.

The scattering term $C(f)$ in Eq. (2.6) evaluates the variation of the phonon distribution function due to the phonon scattering processes. The full expression of the scattering term is extremely complex because of the nonlinear nature of such processes. One common simplification is Callaway's dual relaxation model [40], which assumes that $\mathbf{N}$ and $\mathbf{R}$ processes restore the phonon distribution function separately to a displaced Planck distribution and a Planck distribution, respectively. Callaway's dual relaxation model is widely applied in studying the classical phonon hydrodynamics in low-temperature dielectric crystals [7, 9], where the $\mathbf{N}$ process dominates over the $\mathbf{R}$ process. When we focus on non-Fourier heat conduction at ordinary temperatures where the $\mathbf{N}$ process is negligible except for the very special carbon materials [10, 19], the single-mode relaxation time approximation (SMRTA) is extensively adopted for the phonon scattering term, $C(f) = -\frac{f - f_R^{eq}}{\tau_R}$, where the equilibrium distribution function for the $\mathbf{R}$ process is the Planck distribution: $f_R^{eq} = \{\exp[\hbar\omega/k_B T] - 1\}^{-1}$. At the steady state under small temperature gradient where Fourier's law is valid, the phonon distribution function in the non-equilibrium state slightly differs from that in the equilibrium state so that the phonon scattering process can be simplified as the phonon relaxation time τ_R. Using SMRTA approximation, Eq. (2.6) could be written in the linearized form as [9],

$$f \approx f_R^{eq} - \tau \frac{\partial f_R^{eq}}{\partial T} v_g \cdot \nabla T. \tag{2.7}$$

Substituting Eq. (2.7) into the microscopic definition of the heat flux as follows:

$$\boldsymbol{J} = \iint v_g \hbar\omega f \frac{D(\omega)}{4\pi} d\Omega d\omega, \tag{2.8}$$

and comparing it with Fourier's law gives rise to

$$\kappa = \iint \tau \hbar\omega \frac{D(\omega)}{4\pi} \frac{\partial f_R^{eq}}{\partial T} v_g v_g d\Omega d\omega, \tag{2.9}$$

where $D(\omega)$ is the phonon density of state function. The spectral specific heat capacity is defined as $C_\omega = \hbar\omega D(\omega) \frac{\partial f_R^{eq}}{\partial T}$. Under isotropic approximation in dielectric crystals, thermal conductivity is rewritten as,

$$\kappa = \frac{1}{3} \int C_\omega v_g^2(\omega) \tau(\omega) d\omega. \tag{2.10}$$

Generally the phonon mean fee path Λ, relaxation time τ, and group velocity v_g are frequency-dependent, with the relationship $\Lambda(\omega) = v_g(\omega)\tau(\omega)$. When applying the average quantities of Λ, τ, and v_g, the phonon thermal conductivity of the isotropic material is simplified to Eq. (2.5). According to the Debye model, the phonon contributions to the thermal conductivity can be expressed by following Callaway's equation [40, 41],

$$\kappa = \frac{k_B^4}{2\pi^2 v_g} \left(\frac{T}{\hbar}\right)^3 \int\limits_0^{\theta_D/T} \frac{\tau_C \beta^4 e^\beta}{(e^\beta - 1)^2} d\beta, \tag{2.11}$$

where k_B is the Boltzmann constant, $\hbar$ is Plank's constant, θ_D is the Debye temperature, v_g is the average phonon group velocity, τ_C is the combined relaxation time, and $\beta = \hbar\omega/k_B T$ is a dimensionless quantity. Callaway's approximation considerably simplifies the collision term and is a milestone in the development of the phonon heat transport model.

To solve the phonon Boltzmann equation, one must know the phonon dispersion relation, the group velocity, and the phonon relaxation time. The process starts with the computation of the interatomic force constants based on density functional theory [42] or force-field method [43]. Next, the dynamical matrices are used to solve the BTE with Callaway's approximation. However, accurately calculating the phonon scattering ratio requires pre-knowledge of the physical processes of phonon scatterings. Four different kinds of phonon scattering processes usually occur, namely, the three-phonon Umklapp scattering, phonon-mass impurity scattering, phonon-boundary scattering, and electron-phonon scattering. Based on Matthiessen's rule and without considering the coupling relations among these phonon scattering processes, the combined relaxation time, τ_C, can be written as

$$\tau_C^{-1} = \tau_U^{-1} + \tau_m^{-1} + \tau_b^{-1} + \tau_{ep}^{-1}. \tag{2.11}$$

The relaxation times for **N** and **R** processes usually depend on phonon frequency and crystal temperature. However, unified mathematical expressions for such processes remain lacking. Previous works [40, 41, 44–46] adjusted empirical coefficients in their relaxation time expressions to fit their experimental data of bulk lattice thermal conductivity, which are summarized in Table 2.1. Note that the main difference of the achieved relaxation time expressions lies in the **N** and **U** processes, whereas those of boundary and mass-impurity scatterings have nearly consistent forms. The phonon relaxation time can also be calculated by MD simulations [47], where the anharmonicity is modeled explicitly to all orders.

2.2.3   Molecular Dynamics Simulations

MD simulation is a classical approach that can model the dynamics of each particle in a system of interest based on Newton's second law and the empirical force field. The basics of molecular dynamics are discussed in textbooks [48, 49]. The thermal transport properties of the system can be calculated after the motion trajectory of the particles is obtained through the simulation. The MD method based on the force field does not rely on too many approximations and assumptions on the detailed phonon processes, which are necessities in phonon BTE solutions. Such a method

Table 2.1 Relaxation time expressions for **N** and **R** processes

Ref.	τ_N^{-1}	τ_U^{-1}	τ_b^{-1}	τ_m^{-1}
Callaway [40]	$B_1\omega^2 T^3$	$B_2\omega^2 T^3$	v_g/L	$A\omega^4$
Mingo [41]	Neglected	$B\omega^2 Te^{-C/T}$	v_g/FL	$A\omega^4$
Morelli [44]	$B_{1,i}\omega^2 Te^{-\theta_{D,i}/3T}$	$B_{2,l}\omega^2 T^3, B_{2,t}\omega T^4$	$v_{g,i}/L_{eff}$	$A_i\omega^4$
Ward [45]	$B_1\omega^2 T[1 - e^{-3T/\theta_D}]$	$B_2\omega^4 T[1 - e^{-3T/\theta_D}]$	–	–
Maldovan [46]	Neglected	$B_i\omega^2 Te^{-C/T}$	–	$A\omega^4$

The subscripts m, b denote mass impurity and boundary respectively; The subscript $i=l$, t denotes longitudinal and transverse modes respectively; F means the geometrical factor with the geometrical characteristic dimension L; A, B, and C are empirical coefficients obtained through fitting the experimental data; θ_D is the Debye temperature. Readapted with permission from Ref. [9], copyright (2015) by Elsevier

has numerous advantages, such as the ability to simulate large complex structures [34] while considering atomic level details, including defects [50, 51], strains [52], and surfaces [23, 53], and accounting for anharmonicity to all orders. The MD method can also provide detailed mode-dependent information of phonon, such as the phonon dispersion through the velocity autocorrelation function [54] and the phonon lifetime using the normal mode analysis [55] or spectral energy density analysis [56]. The interaction of phonon waves with interfaces and defects have also been simulated through the wave-packet method [57]. Two different techniques are frequently used in practice to calculate phonon thermal conductivity of bulk- and nano-structured materials, namely, equilibrium MD (EMD) and non-equilibrium MD (NEMD). The EMD method has higher computational cost than the NEMD method. However, the former avoids problems such as finite size and boundary effects, which are inherent in the latter [58]. When the proper limits of extensive durations and large system sizes are carefully considered, the phonon thermal conductivity calculated from the two MD methods are in fact consistent with each other [29], as expected from linear response theory.

2.2.3.1 EMD Method

The EMD method is also referred to as the Green-Kubo method, which is based on the fluctuation-dissipation theorem in classical statistical thermodynamics [59]. In the Green-Kubo scheme, the thermsal conductivity is related to the time needed for the heat current fluctuations to dissipate, which is expressed as,

$$\kappa_{\mu\nu} = \frac{1}{Vk_{\rm B}T^2}\int_0^\infty \langle J_\mu(0) \cdot J_\nu(t)\rangle dt, \qquad (2.12)$$

where V is the volume, k_B is the Boltzmann constant, T is the temperature, J_μ is the μ component of the heat current $J(t)$, and $\kappa_{\mu\nu}$ is an element of the thermal conductivity tensor that is reduced to a scalar in a material with cubic isotropy. The term inside the angle brackets $\langle J_\mu(0) \cdot J_\nu(t) \rangle$ represents the heat current autocorrelation function (HCACF). The thermal conductivity along the three Cartesian directions ($\zeta = x, y, z$) can be calculated by setting $\mu=\nu=\zeta$. The ζ component of the thermal conductivity, κ_ζ, is expressed by

$$\kappa_\zeta = \frac{1}{Vk_BT^2} \int_0^\infty \langle J_\zeta(0) \cdot J_\zeta(t) \rangle dt. \tag{2.13}$$

An important issue associated with the Green-Kubo method is the precise definition of the local energy needed to evaluate the heat current. The heat current $J(t)$ is given by

$$J(t) = \frac{d}{dt} \sum_i E_i(t) r_i(t), \tag{2.14}$$

where $r_i(t)$ is the time-dependent atomic coordinate of atom i and $E_i(t)$ is the microscopic total energy of atom i. For the pair potential, such as the Lennard-Jones potential, wherein total potential energy is written in terms of the pairwise interaction $\varphi(r_{ij})$, the site energy E_i can be expressed as

$$E_i = \frac{1}{2}m_i v_i^2 + \frac{1}{2} \sum_j \varphi(r_{ij}). \tag{2.15}$$

Furthermore, the thermal current $J(t)$ can easily be obtained using

$$J(t) = \sum_i E_i v_i + \frac{1}{2} \sum_{i \neq j} r_{ij}(f_{ij} \cdot v_i). \tag{2.16}$$

where v_i is the velocity vector of atom i, $r_{ij}(= r_i - r_j)$ is the relative position vector of atoms i and j, and f_{ij} is the force on atom i from the pair potential with its neighbor j.

When the interaction involves a many-body nature (e.g., Stillinger-Weber potential [60] and Tersoff potential [61]), the potential energy distribution between particles is not unique, and in fact can be quite arbitrary [62]. In this case, $J(t)$ can be derived from the following general formalism

$$J(t) = \sum_i E_i v_i + \sum_{j \neq i} r_{ji} \left(\frac{\partial E_j}{\partial r_i} \cdot v_i \right) = \sum_i E_i v_i + \sum_{j \neq i} r_{ij} \left(\frac{\partial E_i}{\partial r_j} \cdot v_j \right) \tag{2.17}$$

Fig. 2.5 Flow chart of the EMD Green-Kubo method. NVT denotes the constant-temperature, constant-volume ensemble whereas NVE indicates the constant-energy, constant-volume ensemble

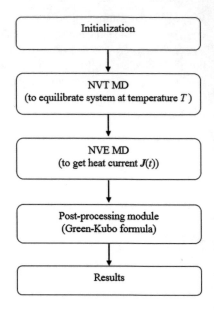

Equation (2.17) provides a clear method of calculating the thermal conductivity in a model system, given that all the dynamic variables in the said equation are explicitly known in an MD simulation.

The ζ component of the thermal conductivity, κ_ζ, is calculated by discretizing the right-hand side of Eq. (2.13) in the EMD timestep Δt as

$$\kappa_\zeta = \frac{\Delta t}{V k_{\mathrm{B}} T^2} \sum_{m=1}^{M} \frac{1}{N_s - m} \sum_{n=1}^{N_s - m} J_\zeta(m+n) \cdot J_\zeta(n), \qquad (2.18)$$

where N_s is the total number of heat current samples after equilibration, M is the number of steps over which the time average is calculated, and $J_\zeta(m+n)$ is the heat current at MD time step $m+n$. Figure 2.5 shows the flow chart of EMD simulation with the Green-Kubo scheme.

Equation (2.13) shows that the temporal decay in the HCACF represents the time scale of the thermal transport, and the thermal conductivity is proportional to the integral of HCACF. In principle, the upper limit of the HCACF integral time in Eq. (2.13) is infinite, whereas the integration time in MD simulations is finite. Thus, as long as one sets an integral time upper limit that is longer than the time taken by the current-current correlations to converge to zero, the results are meaningful. However, the accuracy of HCACF decreases for small ensemble sizes (corresponding to a short correlation time), and numerical noise may also contaminate HCACF when it decays to a small value [63], especially in the simulation of nanostructures, such as nanowires and nanoclusters.

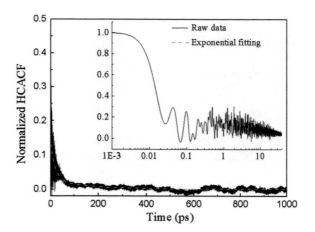

Fig. 2.6 Curves for the normalized HCACF (solid line) of bulk silicon at 300 K and the corresponding double exponential decay fitting (dash line), respectively. The inset is a plot using the logarithmic scale. Reprinted with permission from Ref. [32] copyright (2011) by EPLA

Figure 2.6 shows the time dependence of the HCACF normalized by its zero-time value at 300 K for bulk silicon (512-atom cell). The HCACF decays rapidly in the beginning, followed by a much slower decay. The rapid decay corresponds to the contribution of fast phonons to thermal conductivity, whereas the slower decay corresponds to the contribution of slow phonons, which predominates in thermal conductivity. This two-stage decaying characteristic of the HCACF has been found in the study of various materials. In the case of perfect crystalline silicon, a much longer simulation time (typically on the order of nanoseconds) is needed to obtain barely convergent results of the correlation integral. However, fluctuations in the HCACF during extended correlation times can cause a drift in its integral [64].

Che et al. [65] first used a double exponential decay function to shorten the MD simulation time, avoid cut-off artifacts, and fit the HCACF function, as follows:

$$\langle J_\zeta(0)J_\zeta(t)\rangle = A_o e^{-t/\tau_o} + A_a e^{-t/\tau_a}, \tag{2.19}$$

where subscripts o and a denote fast optical mode and slow acoustic mode, respectively. Substituting Eq. (2.19) into Eq. (2.13), the integrated thermal conductivity can be written as

$$\kappa_\zeta(\tau_m) = \frac{1}{V k_B T^2}\left[A_o \tau_o(1 - e^{-\tau_m/\tau_o}) + A_a \tau_a(1 - e^{-\tau_m/\tau_a})\right]. \tag{2.20}$$

The fitting parameters A_o, τ_o, A_a, and τ_a are then used to calculate thermal conductivity at $\tau_m \to \infty$. Therefore,

$$\kappa_\zeta(\tau_m \to \infty) = \frac{1}{V k_B T^2}(A_o \tau_o + A_a \tau_a). \tag{2.21}$$

Fig. 2.7 Schematic
illustrations for **a** NEMD
and **b** NEGF simulation
setups of the
one-dimensional transport
system, respectively

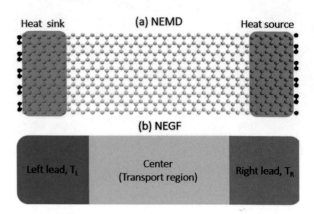

2.2.3.2   NEMD Method

NEMD simulation is widely used to characterize thermal conductivity. As shown in Fig. 2.7a, a temperature gradient can be built in the heat flux direction by using two heat reservoirs [66] at the opposite ends of the system. This approach is analogous to a real experimental measurement and is known as the direct NEMD (d-NEMD) method [20]. Alternatively, in the reverse NEMD (r-NEMD) method, a heat flux can be directly imposed to the system by adding/subtracting kinetic energies and subsequently determines the induced temperature gradient [20, 67]. In both NEMD methods, the simulation must be run long enough to reach the steady state where the heat flux and temperature profile are constant. Once the system of interest has reached a non-equilibrium steady state, the temperature gradient can be calculated from the linear fit line of the temperature profile, and the heat flux can be computed according to the energy injected into/extracted from the heat source/sink across the unit area per unit time. Thus, thermal conductivity can be ascertained based on Fourier's law of heat conduction.

The Müller-Plathe method is a famous algorithm for implementing the r-NEMD approach [68]. Apart from its very simple implementation, the Müller-Plathe method offers several advantages, such as compatibility with periodic boundary conditions, conservation of total energy, and conservation of total linear momentum. The basic idea of the Müller-Plathe method is to divide the simulation system into N slabs along the heat flux direction (e.g., x direction). As shown in Fig. 2.8a, two slabs in the left and right edges are considered cold slabs, and the middle slab is regarded as the hot slab. Note that the thickness of the hot region doubles that of the cold region to ensure that the number of atoms in the hot region is the same as that in the two cold regions. A heat flux J_x is created by exchanging the momentum between the hottest atom velocity in the cold (sink) region and the coldest atom velocity in the hot (source) region for every swap time interval, which induces a temperature gradient in the system of interest. After N_{swap} exchanges are performed, the heat flux J_x is given by

Fig. 2.8 **a** Schematic of the simulation model for the Müller-Plathe method. **b** Typical temperature profile in a nanoribbon along the x direction at 300 K. Reprinted with permission from Ref. [33], copyright (2018) by the Chinese Physical Society and IOP Publishing

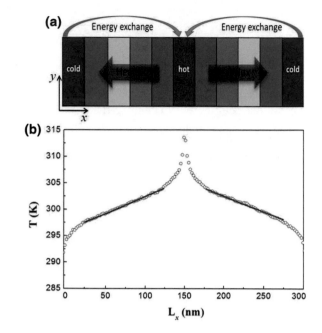

$$J_x = \frac{\sum_{N_{swap}} \frac{1}{2}m(v_h^2 - v_c^2)}{2\tau_{swap}A}, \quad (2.22)$$

where τ_{swap} is the time period of the exchange simulation, m is the atom mass, and v_h and v_c are the velocities chosen from the hot and cold regions at each swap time interval, respectively. A is the cross-sectional area of the simulation system perpendicular to the heat flux direction, and the factor "2" arises from the fact that the heat can flow from the hot slab to the cold slab in two directions. In the NVE ensemble, the temperature in each slab is calculated based on the velocities of atoms inside that slab as

$$T_i = \frac{1}{3N_i k_B} \sum_{j=1}^{N_i} m_j v_j^2, \quad (2.23)$$

where N_i is the number of atoms in slab i. m_j and v_j are the mass and velocity of atom j in slab i, respectively. Figure 2.8b displays a typical time-averaged temperature profile as a function of the slab position across the nanoribbon. The temperature distributions near the hot and cold slabs are usually nonlinear as the velocities of atoms are exchanged to generate the temperature gradient, $\delta T / \delta x$. $\delta T / \delta x$ is obtained from the linear parts of the temperature distribution in the middle. Based on Fourier's law of heat conduction, the thermal conductivity κ is calculated by

$$\kappa = \frac{\langle J_x \rangle}{\langle \delta T / \delta x \rangle},$$ (2.24)

where the brackets denote a statistical time average. The ratio can produce a reasonable κ value, provided that the system is fully equilibrated and the simulation is sampled over a sufficiently long time.

Note that both the NEMD and EMD methods, which are based on the classical Newton's law, can not take quantum effect into consideration. Therefore, the classical MD methods are unable to reproduce quantum thermal conductance at low temperatures and in quantum structures. Quantum correction [69] and quantum approximation [70] have been proposed to reduce this limitation in classical MD simulations.

2.2.4 Non-equilibrium Green's Function Method

The non-equilibrium Green's function (NEGF) formalism is a well-developed quantum theory to exactly deal with *ballistic* thermal transport [35]. Also known as the atomistic Green's function method [36], the NEGF is based on the quantization of lattice dynamics and scattering theories. The NEGF method was initially developed to handle electrical transport [71, 72], but its application for the investigation of the thermal transport properties of nanomaterials has become increasing popular [37, 73]. Details of the NEGF methodology are available in the review articles [36, 74].

As shown in Fig. 2.7b, the system of interest at the center is connected to left (L) and right (R) semi-infinite thermal leads at different temperatures T_L and T_R, respectively. In the *ballistic* thermal transport limit (which means that the phonons do not lose or gain energy during travel), the heat flux flowing from the left to the right lead is given by the Landauer formula [35],

$$J = \int_0^\infty \frac{d\omega}{2\pi} \hbar\omega\zeta(\omega)[f_L(\omega) - f_R(\omega)],$$ (2.25)

where $f_{L(R)}(\omega) = \{\exp[\hbar\omega/k_B T_{L(R)}] - 1\}^{-1}$ is the Bose-Einstein distribution of the left (right) regime, $\zeta(\omega)$ is the transmission coefficient, and $\hbar\omega$ represents the energy of the phonons. Given the limit of a very small temperature difference between the two leads, thermal conductance can be written as

$$\sigma = \int_0^\infty \frac{d\omega}{2\pi} \hbar\omega\zeta(\omega)\frac{\partial f(\omega)}{\partial T}.$$ (2.26)

The phonon transmission coefficient $\zeta(\omega)$ can be calculated starting from phonon wave-package dynamics [57] or by using the Green's function technique [75]. For a non-interacting harmonic system, the Hamiltonian is $H_0 = 2^{-1}(\dot{u}^{\mathrm{T}}\dot{u} + u^{\mathrm{T}}Ku)$, where u is a column vector consisting of all the displacement variables in a region and $\dot{u}$ is the corresponding conjugate momentum. The superscript T stands for matrix transpose. The retarded Green function in the frequency domain is given by $G^{\mathrm{r}}[\omega] = [(\omega + i\eta)^2 I - K]^{-1}$ where ω is the vibrational frequency of phonons, I is the identity matrix, and $\eta \to 0^+$. K is the force constant matrix, derived by the empirical force field or density functional theory, and calculations with the former will be much faster than those with the latter. Then, phonon transmission can be calculated as [76]

$$\zeta(\omega) = \mathrm{Tr}[\Gamma_{\mathrm{L}}(\omega)G^{\mathrm{a}}(\omega)\Gamma_{\mathrm{R}}(\omega)G^{\mathrm{r}}(\omega)], \qquad (2.17)$$

where Tr means taking the trace, $G^{\mathrm{r,a}}(\omega)$ is the retarded or advanced Green function for the central region related by $G^{\mathrm{r}}(\omega) = (G^{\mathrm{a}}(\omega))^{\dagger}$ and $\Gamma_{\mathrm{L,R}}(\omega)$ describes the coupling between the leads and the center. When nonlinear interaction (i.e., anharmonity) at the central part is further considered, the NEGF provides a more complicated expression to evaluate thermal transport [21]. Detailed algorithms for the calculation of all the Green's functions are available in [77].

Note that the NEGF is particularly strong in handling low-temperature thermal transport where the quantum effect is dominant, but its ability to model systems with many atoms is limited. For simple systems like silicon nanowires, the NEGF method has been used extensively to study the effects of various phenomena on thermal transport, such as anisotropy [78], surface roughness [79], and structural and substitutional defects [80, 81]. The NEGF method has also been applied for complex systems, such as the thermoelectric properties of hybrid nanostructures [82], and the thermal conductivity of graphene [83] and the ZnO/ZnS interface [84]. Furthermore, technological improvements based on NEGF formalism are continuously developing to handle phonon-phonon scattering [35], electron-phonon scattering [85], and interfacial thermal transport [86].

2.3 Experimental Methods for Determining Nanoscale Thermal Conductivity

Experimental studies on thermal transport in low-dimensional materials are relatively rare due to the difficulty of simultaneous measurements of heat flux and temperature distribution at the nanoscale. With the advance of nano-fabrication technologies, heating nanoscale materials and measuring temperature gradients can be performed simultaneously, although with great challenges. Thus far, various techniques have been invented to determine the intrinsic thermal conductivity of low-dimensional materials, including the thermal bridge method [87], optothermal Raman technique [88], three omega technique [89], and time-domain thermoreflectance method [90]. Successful measurement of nanoscale thermal conductivity has helped to understand phonon transport in nanomaterials from both a fundamental and a practical perspective [21]. In this section, we specifically review popular technologies for assessing the thermal conductivity of silicon nanomaterials: the thermal bridge method, optothermal Raman technique, and 3ω method.

2.3.1 Thermal Bridge Method

In 2001, Kim et al. [91] introduced an original measurement, the so-called thermal bridge method, which entails suspending a sample between two membranes serving as a heat source and a heat sink [21]. A representative example is the approach later developed by Shi et al. [87]. A suspended micro-electro-thermal system (METS) device and a nano-manipulation system were employed to suspend low-dimensional materials and detect temperature changes at the nanoscale, respectively. The METS device is mass-fabricated using a standard wafer-stage nanofabricating process. This technique has been extensively utilized to evaluate the thermal conductivity of one-dimensional nanotubes [87, 91, 92] and nanowires [93–95], and two-dimensional suspended [96] and supported [97] graphene. Numerous studies also applied the thermal bridge method on thermoelectric materials because of the demand for reduced thermal conductivity in nanostructures [98–100].

The measurement using a suspended micro-device is based on a steady-state heat transfer model. In the thermal bridge method, thermal transport measurements were conducted using a METS device in vacuum ($<10^{-5}$ mbar). As described previously in [21, 87, 92], the METS device consists of a 300 nm thick SiN_x layer as mechanical support and an integrated Pt resistive loop that serves as both heater and temperature sensor. Each of the heater/sensor islands is supported by Pt leads on SiN_x beams from the substrate. A nanomanipulator in a scanning electron microscope (SEM) was used to pick up and place individual nanowires (or nanoribbons) onto the METS device (Fig. 2.9a). To reduce thermal contact resistance, two ends of the sample were bonded onto the Pt electrodes by Pt-C composite using electron-beam-induced deposition (EBID). A direct current (I_{dc}) is passed through the heater resistor (R_h) for heating, and to increase its temperature (T_h) from the environmental temperature (T_0). The Joule heat with a heating power of $P = I_{dc}^2 R_h$ in the heater gradually dissipates through the Pt/SiN$_x$ beams and the sample connecting them, thus raising the temperature (T_s) in the sensor resistor (R_s). The four-terminal electrical resistance of the Pt loops were obtained using lock-in amplifiers by passing a very low alternating current (I_{ac}) that superimposes on I_{dc}. The heater and sensor temperatures (T_h and T_s) were then obtained based on R_h and R_s, which were calibrated against the substrate temperature (T_0) of the METS. On the basis of the temperature changes and the heating power P at the heater (Fig. 2.9c), the reciprocal of thermal conductance (G) of the sample, R, can be determined through thermal resistance circuit (Fig. 2.9b) analysis as found in [87],

$$R = \frac{1}{G_s} = R_b \left(\frac{\Delta T_h - \Delta T_s}{\Delta T_s} \right), \tag{2.18}$$

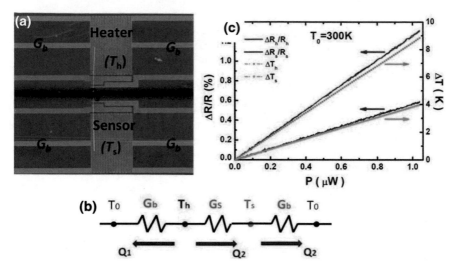

Fig. 2.9 Thermal resistance circuit and measurement details for the METS device. a SEM image of a METS device with an individual nanowire bonded on the heater and sensor for measuring the thermal conductivity, thermopower, and electrical resistance in a single device. b Schematic of the thermal flow circuit. c Resistance change and corresponding temperature change in the heater/sensor. Reprinted with permission from Ref. [21] copyright (2016) by IOP Publishing

where $\Delta T_{h,s} = T_{h,s} - T_0$ and the thermal conductance of the connecting beams is $R_b = (\Delta T_h + \Delta T_s)/P$, which can also be obtained from the thermal conductivity and dimension of the beams.

The measured R is the overall thermal resistance from the heater to the sensor, which includes the contact thermal resistance at the two ends of the sample, R_C, and the intrinsic thermal resistance of the suspended sample, $R_{int} = L/A\kappa$, with the cross-section area A and sample length L. Thus, the thermal conductivity of the sample can be determined by

$$\kappa = \frac{L}{A(R - R_C)}. \tag{2.19}$$

Note that the main challenge for the thermal bridge method lies in the uncertainty contribution of R_C [25]. One proposed solution entails reducing the R_C in the experiment such that R_C can be lowered to the negligible level or below [101]. However, if R_C is more significant (and thus non-negligible), then one can repeat measurements with two or more samples with different sizes to extract the R_C and κ of the sample [102]. Recently, based on the previous METS device, an electron-beam heating technique [103] has been developed for the spatially resolved measurement of cumulative thermal resistance. A focused electron-beam with 2–3 nm diameter inside SEM was used to induce localized heating along the sample, and both the heater and sensor acted as temperature sensors (Fig. 2.10). By scanning the electron beam from the

Fig. 2.10 a Electron-beam heating technique where a focused electron beam was used as a local heating source and scanned along the length of the suspended nanowire. **b** Equivalent thermal resistance circuit showing the cumulative thermal resistance, R_i, from the left island to the heating spot and the temperature rise of the left and right sensors (ΔT_L and ΔT_R). By solving for the heat flow on the left and right suspended pads, length-dependent thermal conductivity can be determined. Reprinted with permission from Ref. [103], copyright (2014) by the American Chemical Society

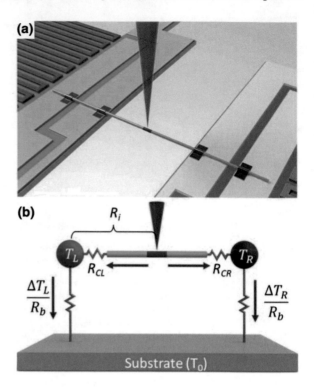

heater source to the contact area of the nanowire, length-dependent thermal conductivity can be obtained, which is a powerful addition to the current technology. The electron-beam scanning along the sample rarely affects its thermal resistance. After the electron-beam scanning, repeated measurements of total thermal resistance R produce essentially identical values.

2.3.2 Optothermal Raman Technique

The Raman spectroscopy-based method is one of the most commonly used techniques for measuring the thermal conductivity of low-dimensional materials. The first experimental study of heat conduction in graphene was made possible by developing an optothermal Raman technique [3, 88]. The heating power ΔP was provided by a laser light focused on a suspended graphene layer connected to heat sinks at its ends (Fig. 2.11a). Temperature rise (ΔT) in response to ΔP was measured by the temperature-induced Raman shift. Initially, the calibration of the spectral position of the G peak with T (i.e., calibration curve $\omega_G(T)$ in Fig. 2.11b) was performed by changing the sample temperature while using very low laser power to avoid local heating, which allows for the conversion of a Raman spectrometer into a Raman ther-

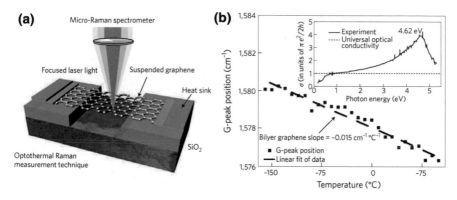

Fig. 2.11 **a** Schematic of the sample geometry for thermal conductivity measurement using a micro-Raman spectrometer. **b** The measured relationship between the temperature and the G peak shifts of the Raman signals for a bilayer graphene. The Raman G peak shift frequency (ω_G) changes linearly with the temperature (T) according to the coefficient $\chi_G = \delta\omega_G/\delta T$. The inset in **b** shows that the optical absorption in graphene is a function of the light wavelength. Reprinted with permission from Ref. [3], copyright (2011) by Springer Nature

mometer. Then, during thermal conductivity measurements, the suspended graphene layer is heated by increasing laser power. Local ΔT in graphene is determined by $\Delta T = \Delta\omega_G/\chi_G$, where χ_G is the T coefficient of the G peak. The amount of heat dissipated in graphene can be obtained either by measuring the integrated Raman intensity of the G peak [88] or using a detector placed under the graphene layer [104]. A correlation between ΔT and ΔP for a graphene sample with a given geometry can provide a thermal conductivity value through solving the heat-diffusion equation. For example, heat is assumed to spread from the center in two opposite directions when the width of the trench is far greater than the laser spot size for a suspended graphene cross trench. In this case, the thermal conductivity of the suspended graphene can be obtained by [88], such that

$$\kappa = \left(\frac{L}{4S}\right)\left(\frac{\Delta P}{\Delta T}\right),\tag{2.20}$$

where L is the length of the sample suspended between two stripes, T is the absolute temperature, and S is the cross-section area of the graphene perpendicular to the direction of heat flux.

The optothermal Raman technique for measuring κ is a direct steady-state method. Given its easy accessibility, the optothermal Raman technique has been extended to other two-dimensional materials (such as supported graphene [105], isotopically modified graphene [106] and MoS_2 [107], silicon nanofilm [108]), and one-dimensional materials (such as silicon nanowires [109] and nanotubes [110, 111]). Note that the accuracy and precision of the Raman-derived thermal conductivity depend on the following [112]: assumptions within the analytical model used in the deduction of thermal conductivity, uncertainty in the quantification of heat flux and

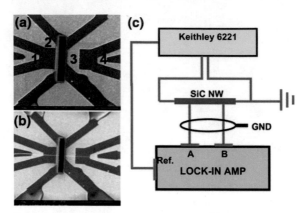

Fig. 2.12 Four-point-probe 3ω experiment. **a** a prepatterned four-point-probe, **b** a single NW placed by the nanomanipulator, and **c** schematic of the measurement setup. The 3ω-signal of a sample (for example, β-SiC NW) is measured by using a lock-in amplifier and an alternating current. Source (Keithley 6221). Readapted with permission from Ref. [114], copyright (2010) by IOP Publishing

temperature arising from the conditions of the experiment (defects, contaminations, and near-field or multiple reflection effects) as well as the evolution of thermomechanical stress during testing.

2.3.3 Three Omega Technique

As a frequency-domain transient technique, the 3ω method is widely used to measure the thermal properties of both bulk materials and thin films after it was first introduced in 1990 by Cahill and coworkers [113]. Originating from the celebrated method reported by Cahill for thin film measurements, the four-point-probe 3ω method was developed to determine the thermal conductivity of one-dimensional materials including nanowires [89, 114, 115] and nanoribbons [116]. The four-point-probe 3ω method is based on the fact that the third harmonic amplitude as a response to AC current applied at the fundamental frequency ω can be expressed in terms of thermal conductivity. With this method, the effect of contact resistance, which normally occurs in the conventional two-point-probing method, can be eliminated.

For the nanowire measurements, the sample is suspended with four electrical contacts, as illustrated in Figs. 2.12a, b. A sinusoidal current at 1ω angular frequency is supplied to the sample and induces a temperature modulation at the 2ω frequency because of Joule heating. In turn, the 2ω temperature fluctuation generates a 3ω component in the voltage drop measured across the suspended sample. Measured with a lock-in amplifier (Fig. 2.12c), the *rms* amplitude of the 3ω voltage component at the low frequency limit is given by [114],

$$V_{3\omega,rms} \approx \frac{\sqrt{2}I_0^3 RR'L}{\pi^4 \kappa S}, \tag{2.21}$$

where L, $R[= R_0 + R'(T - T_0)]$, and S are the length, electrical resistance, and cross-sectional area of the nanowire, respectively. I_0 is the current amplitude. The resistance change with temperature at room temperature, $R' = (\partial R/\partial T)_{T_0}$, should be pre-measured, and κ is the thermal conductivity of the nanowire. To obtain the accurate third harmonic response of the system (within 1.2%), the condition of $I_0^2 R'L/n^2\pi^2\kappa S \ll 1$ should be satisfied [89]. The initial and boundary conditions for the temperature are room temperature (T_0) at time $t = 0$ and at the contacts between the nanowire and the metal electrodes, respectively.

2.4   Concluding Remarks

In this chapter, we briefly discussed the fundamental phonon thermal conduction at the nanoscale and outlined various theoretical methods and experimental techniques commonly used in the literature to determine the phonon thermal conductivity in low-dimensional materials. With the development of novel theoretical and experimental approaches, the past few decades witnessed significant advancements in the understanding and tuning of the thermal properties of various nanomaterials, including nanoparticles, nanowires, and nanosheets.

Theoretical studies are expected to play a significant role in comprehending heat transport at the nanoscale, but they must be supported by experimental data. Reducing the size of nanomaterials is becoming possible through different experimental techniques, such as electrodeposition [117] or oxide–assisted growth [118] for silicon nanowires and mechanical exfoliation [119] or chemical vapor deposition [120] for graphene. However, important challenges and debates about both experimental and computational techniques deserve further investigation. Understanding and controlling the thermal transport properties in general and thermoelectric materials in particular, are crucial for different fields, including nanoenergy and thermal management.

As discussed, many computational studies have been conducted on phonon transport in low-dimensional nanomaterials. The applications of different level methods have limitations at present. Choosing an appropriate method to study different problems at the nanoscale is crucial. In the framework of classical mechanics, Newton's equation is solved as the foundation of the MD simulation method, whereas in the quantum mechanical framework, the Schrödinger equation is solved that corresponds to the first-principles method and NEGF method [9]. Conventional first-principles methods theoretically offer high accuracy but cannot handle a large system, whereas the empirical potentials used in classical MD often lack transferability or accuracy because of overlooking electronic contributions [23]. Therefore, developing improved theory and methods to rapidly and accurately calculate the thermal conductivity of nanostructures is necessary.

Special attention should be paid to experimental observations, despite the ongoing heated debate, because limited experimental techniques have been available for determining nanoscale thermal conductivity until now. The primary challenge for the thermal bridge method lies in the uncertainty contribution from the contact resistance at the two ends of the sample. To overcome this challenge, efforts should be made in future to directly measure thermal contact resistance. Optothermal Raman spectroscopy is sensitive to the temperature dependence of the phonon peak shift as well as the laser energy absorption in nanomaterials such that any incorrect determination of these two factors may cause large errors. In addition, the transient 3ω method should be developed further to measure the thermal conductivity of two-dimensional nanostructures, such as graphene and silicene, although it has achieved fruitful results for one-dimensional nanowires and nanotubes. Given the discussed experimental challenges and the effects from size, surface, and defects, differences occur in the results of the thermal conductivity measurements for nanomaterials among different research groups at present.

References

1. Fourier, J.: The Analytical Theory of Heat. Dover, New York (1955). (Reprint of Fourier's 1822 monograph)
2. Ghosh, S., Bao, W., Nika, D.L., Subrina, S., Pokatilov, E.P., Lau, C.N., Balandin, A.A.: Dimensional crossover of thermal transport in few-layer graphene. Nat. Mater. 9(7), 555–558 (2010). https://doi.org/10.1038/nmat2753
3. Balandin, A.A.: Thermal properties of graphene and nanostructured carbon materials. Nat. Mater. 10(8), 569–581 (2011). https://doi.org/10.1038/nmat3064
4. Glassbrenner, C.J., Slack, G.A.: Thermal conductivity of silicon and germanium from 3°K to the melting point. Phys. Rev. 134, A1058 (1964). https://doi.org/10.1103/PhysRev.134.A1058
5. Tiwari, M.D., Agrawal, B.K.: Analysis of the lattice thermal conductivity of germanium. Phys. Rev. B 4, 3527 (1971). https://doi.org/10.1103/PhysRevB.4.3527
6. Chen, G.: Particularities of heat conduction in nanostructures. J. Nanopart. Res. 2(2), 199–204 (2000). https://doi.org/10.1023/A:1010003718481
7. Zeng, Y.J., Liu, Y.Y., Zhou, W.X., Chen, K.Q.: Nanoscale thermal transport: theoretical method and application. Chin. Phys. B 27(3), 036304 (2018). https://doi.org/10.1088/1674-1056/27/3/036304
8. Lindsay, L., Broido, D.A., Mingo, N.: Flexural phonons and thermal transport in graphene. Phys. Rev. B 82, 115427 (2010). https://doi.org/10.1103/PhysRevB.82.115427
9. Guo, Y.Y., Wang, M.R.: Phonon hydrodynamics and its applications in nanoscale heat transport. Phys. Rep. 595, 1–44 (2015). https://doi.org/10.1016/j.physrep.2015.07.003
10. Lee, S., Broido, D., Esfarjani, K., Chen, G.: Hydrodynamic phonon transport in suspended graphene. Nat. Commun. 6, 6290 (2015). https://doi.org/10.1038/ncomms7290
11. Rego, L.G., Kirczenow, G.: Quantized thermal conductance of dielectric quantum wires. Phys. Rev. Lett. 81(1), 232 (1998). https://doi.org/10.1103/PhysRevLett.81.232
12. Wang, J., Wang, J.S.: Carbon nanotube thermal transport: ballistic to diffusive. Appl. Phys. Lett. 88(11), 111909 (2006). https://doi.org/10.1063/1.2185727
13. Vineis, C.J., Shakouri, A., Majumdar, A., Kanatzidis, M.G.: Nanostructured thermoelectrics: big efficiency gains from small features. Adv. Mater. 22(36), 3970–3980 (2010). https://doi.org/10.1002/adma.201000839

14. Chen, Y.R., Jeng, M.S., Chou, Y.W., Yang, C.C.: Molecular dynamics simulation of the thermal conductivities of Si nanowires with various roughnesses. Comput. Mater. Sci. **50**(6), 1932–1936 (2011). https://doi.org/10.1016/j.commatsci.2011.01.047
15. Cahill, D.G., Braun, P.V., Chen, G., Clarke, D.R., Fan, S., Goodson, K.E., Keblinski, P., King, W.P., Mahan, G.D., Majumdar, A., Maris, H.J.: Nanoscale thermal transport. II. 2003–2012. Appl. Phys. Rev. **1**(1), 011305 (2014). https://doi.org/10.1063/1.4832615
16. Chen, G.: Multiscale simulation of phonon and electron thermal transport. Annu. Rev. Heat. Transf. **17**, 1–8 (2014). https://doi.org/10.1615/AnnualRevHeatTransfer.2014011051
17. Alvarez, F.X., Jou, D., Sellitto, A.: Phonon hydrodynamics and phonon–boundary scattering in nanosystems. J. Appl. Phys. **105**, 014317 (2009). https://doi.org/10.1063/1.3056136
18. Muscato, O., Castiglione, T., Coco, A.: Hydrodynamic modeling of electron transport in silicon quantum wires. J. Phys: Conf. Ser. **906**, 012010 (2017). https://doi.org/10.1088/1742-6596/906/1/012010
19. Guo, Y., Wang, M.: Phonon hydrodynamics for nanoscale heat transport at ordinary temperatures. Phys. Rev. B **97**(3), 035421 (2018). https://doi.org/10.1103/PhysRevB.97.035421
20. Zhang, G. (ed.): Nanoscale energy transport and harvesting: A computational study. Pan Stanford, Singapore (2015)
21. Xu, X., Chen, J., Li, B.: Phonon thermal conduction in novel 2D materials. J. Phys.: Condens. Matter **28**, 483001 (2016). https://doi.org/10.1088/0953-8984/28/48/483001
22. Li, N., Ren, J., Wang, L., Zhang, G., Hänggi, P., Li, B.: Phononics: manipulating heat flow with electronic analogs and beyond. Rev. Mod. Phys. **84**, 1045 (2012). https://doi.org/10.11 03/RevModPhys.84.1045
23. Li, H.P., Zhang, R.Q.: Surface effects on the thermal conductivity of silicon nanowires. Chin. Phys. B **27**(3), 036801 (2018). https://doi.org/10.1088/1674-1056/27/3/036801
24. Toberer, E.S., Baranowski, L.L., Dames, C.: Advances in thermal conductivity. Annu. Rev. Mater. Res. **42**, 179–209 (2012). https://doi.org/10.1146/annurev-matsci-070511-155040
25. Gu, X., Wei, Y., Yin, X., Li, B., Yang, R.: Phononic thermal properties of two-dimensional materials. arXiv:1705.06156 (2017). https://arxiv.org/abs/1705.06156
26. Guo, Y.Y., Wang, M.R.: Phonon hydrodynamics: progress, applications and perspectives. Sci. Sin. Phys. Mech. Astron. **47**(7), 070010 (2017). https://doi.org/10.1360/SSPMA2016-00408
27. McGaughey, A.J.H., Kaviany, M.: Quantitative validation of the Boltzmann transport equation phonon thermal conductivity model under the single-mode relaxation time approximation. Phys. Rev. B **69**, 094303 (2004). https://doi.org/10.1103/PhysRevB.69.094303
28. Zou, J., Balandin, A.: Phonon heat conduction in a semiconductor nanowire. J. Appl. Phys. **89**, 2932 (2001). https://doi.org/10.1063/1.1345515
29. Dong, H., Fan, Z., Shi, L., Harju, A., Ala-Nissila, T.: Equivalence of the equilibrium and the nonequilibrium molecular dynamics methods for thermal conductivity calculations: From bulk to nanowire silicon. Phys. Rev. B **97**, 094305 (2018). https://doi.org/10.1103/PhysRev B.97.094305
30. Li, H.P., Zhang, R.Q.: Vacancy-defect-induced diminution of thermal conductivity in silicene. EPL **99**(3), 36001 (2012). https://doi.org/10.1209/0295-5075/99/36001
31. Li, H.P., Zhang, R.Q.: Anomalous effect of hydrogenation on phonon thermal conductivity in thin silicon nanowires. EPL **105**(5), 56003 (2014). https://doi.org/10.1209/0295-5075/10 5/56003
32. Li, H.P., De Sarkar, A., Zhang, R.Q.: Surface-nitrogenation-induced thermal conductivity attenuation in silicon nanowires. EPL **96**(5), 56007 (2011). https://doi.org/10.1209/0295-50 75/96/56007
33. Xu, R.F., Han, K., Li, H.P.: Effect of isotope doping on phonon thermal conductivity of silicene nanoribbons: a molecular dynamics study. Chin. Phys. B **27**(2), 026801 (2018). https://doi.o rg/10.1088/1674-1056/27/2/026801
34. Chen, J., Zhang, G., Li, B.: Tunable thermal conductivity of $Si_{1-x}Ge_x$ nanowires. Appl. Phys. Lett. **95**(7), 073117 (2009). https://doi.org/10.1063/1.3212737
35. Xu, Y., Wang, J.S., Duan, W., Gu, B.L., Li, B.: Nonequilibrium Green's function method for phonon-phonon interactions and ballistic-diffusive thermal transport. Phys. Rev. B **78**(22), 224303 (2008). https://doi.org/10.1103/PhysRevB.78.224303

36. Zhang, W., Fisher, T.S., Mingo, N.: The atomistic Green's function method: an efficient simulation approach for nanoscale phonon transport. Numer. Heat Transf. B **51**, 333–349 (2007). https://doi.org/10.1080/10407790601144755
37. Xie, Z.X., Chen, K.Q., Duan, W.: Thermal transport by phonons in zigzag graphene nanoribbons with structural defects. J. Phys.: Condens. Matter **23**(31), 315302 (2011). https://doi.or g/10.1088/0953-8984/23/31/315302
38. Boltzmann, L.: Weitere Studien über das Wärmegleichgewicht unter Gasmolekülen. In: Kinetische Theorie II. WTB Wissenschaftliche Taschenbücher, vol. 67, pp. 115–225. Vieweg + Teubner Verlag, Wiesbaden (1970). https://doi.org/10.1007/978-3-322-84986-1_3
39. Peierls, R.: Zur kinetischen theorie der wärmeleitung in kristallen. Ann. Phys. **395**(8), 1055–31101 (1929). https://doi.org/10.1002/andp.19293950803
40. Callaway, J.: Model for lattice thermal conductivity at low temperatures. Phys. Rev. **113**, 1046 (1959). https://doi.org/10.1103/PhysRev.113.1046
41. Mingo, N., Yang, L., Li, D., Majumdar, A.: Predicting the thermal conductivity of Si and Ge nanowires. Nano Lett. **3**(12), 1713–1716 (2003). https://doi.org/10.1021/nl034721i
42. Broido, D., Malorny, M., Birner, G., Mingo, N., Stewart, D.: Intrinsic lattice thermal conductivity of semiconductors from first principles. Appl. Phys. Lett. **91**, 231922 (2007). https://d oi.org/10.1063/1.2822891
43. Broido, D.A., Ward, A., Mingo, N.: Lattice thermal conductivity of silicon from empirical interatomic potentials. Phys. Rev. B **72**, 014308 (2005). https://doi.org/10.1103/PhysRevB.7 2.014308
44. Morelli, D.T., Heremans, J.P., Slack, G.A.: Estimation of the isotope effect on the lattice thermal conductivity of group IV and group III–V semiconductors. Phys. Rev. B **66**(19), 195304 (2002). https://doi.org/10.1103/PhysRevB.66.195304
45. Ward, A., Broido, D.A.: Intrinsic phonon relaxation times from first-principles studies of the thermal conductivities of Si and Ge. Phys. Rev. B **81**(8), 085205 (2010). https://doi.org/10.1 103/PhysRevB.81.085205
46. Maldovan, M.: Thermal conductivity of semiconductor nanowires from micro to nano length scales. J. Appl. Phys. **111**(2), 024311 (2012). https://doi.org/10.1063/1.3677973
47. Qiu, B., Ruan, X.: Reduction of spectral phonon relaxation times from suspended to supported graphene. Appl. Phys. Lett. **100**(19), 193101 (2012). https://doi.org/10.1063/1.4712041
48. Rapaport, D.C.: The art of molecular dynamics simulation. Cambridge University Press, Cambridge (2004)
49. Haile, J.M.: Molecular dynamics simulation: elementary methods. John Wiley and Sons, New York (1992)
50. Ebrahimi, S., Azizi, M.: The effect of high concentrations and orientations of stone-wales defects on the thermal conductivity of graphene nanoribbons. Mol. Simulat. **44**(3), 1–7 (2017). https://doi.org/10.1080/08927022.2017.1366654
51. Hu, S., Chen, J., Yang, N., Li, B.: Thermal transport in graphene with defect and doping: phonon modes analysis. Carbon **116**, 139–144 (2017). https://doi.org/10.1016/j.carbon.201 7.01.089
52. Li, X., Maute, K., Dunn, M.L., Yang, R.: Strain effects on the thermal conductivity of nanostructures. Phys. Rev. B **81**(24), 245318 (2010). https://doi.org/10.1103/PhysRevB.81.245318
53. Sun, D.Y., Liu, J.W., Gong, X.G., Liu, Z.F.: Empirical potential for the interaction between molecular hydrogen and graphite. Phys. Rev. B **75**(7), 075424 (2007). https://doi.org/10.110 3/PhysRevB.75.075424
54. Dickey, J.M., Paskin, A.: Computer simulation of the lattice dynamics of solids. Phys. Rev. **188**(3), 1407 (1969). https://doi.org/10.1103/PhysRev.188.1407
55. Feng, T., Qiu, B., Ruan, X.: Anharmonicity and necessity of phonon eigenvectors in the phonon normal mode analysis. J. Appl. Phys. **117**(19), 195102 (2015). https://doi.org/10.10 63/1.4921108
56. Thomas, J.A., Turney, J.E., Iutzi, R.M., Amon, C.H., McGaughey, A.J.: Predicting phonon dispersion relations and lifetimes from the spectral energy density. Phys. Rev. B **81**(8), 081411 (2010). https://doi.org/10.1103/PhysRevB.81.081411

57. Schelling, P.K., Phillpot, S.R., Keblinski, P.: Phonon wave-packet dynamics at semiconductor interfaces by molecular-dynamics simulation. Appl. Phys. Lett. **80**, 2484 (2002). https://doi.org/10.1063/1.1465106

58. Schelling, P.K., Phillpot, S.R., Keblinski, P.: Comparison of atomic-level simulation methods for computing thermal conductivity. Phys. Rev. B **65**, 144306 (2002). https://doi.org/10.1103/PhysRevB.65.144306

59. Kubo, R.: Statistical-mechanical theory of irreversible processes. I. General theory and simple applications to magnetic and conduction problems. J. Phys. Soc. Jpn. **12**(6), 570–586 (1957). https://doi.org/10.1143/JPSJ.12.570

60. Stillinger, F.H., Weber, T.A.: Computer simulation of local order in condensed phases of silicon. Phys. Rev. B **31**, 5262 (1985). https://doi.org/10.1103/PhysRevB.31.5262

61. Tersoff, J.: New empirical approach for the structure and energy of covalent systems. Phys. Rev. B **37**(12), 6991 (1988). https://doi.org/10.1103/PhysRevB.37.6991

62. Li, J., Porter, L., Yip, S.: Atomistic modeling of finite-temperature properties of crystalline β-SiC: II. Thermal conductivity and effects of point defects. J. Nucl. Mater. **255**, 139–152 (1998). https://doi.org/10.1016/S0022-3115(98)00034-8

63. McGaughey, A.J.H., Kaviany, M.: Thermal conductivity decomposition and analysis using molecular dynamics simulations: Part II. Complex silica structures. Int. J. Heat Mass Transf. **47**(8–9), 1799–1816 (2004). https://doi.org/10.1016/j.ijheatmasstransfer.2003.11.009

64. Chen, J., Zhang, G., Li, B.: How to improve the accuracy of equilibrium molecular dynamics for computation of thermal conductivity. Phys. Lett. A **374**(23), 2392–2396 (2010). https://doi.org/10.1016/j.physleta.2010.03.067

65. Che, J., Çağın, T., Deng, W., Goddard III, W.A.: Thermal conductivity of diamond and related materials from molecular dynamics simulations. J. Chem. Phys. **113**(16), 6888–6900 (2000). https://doi.org/10.1063/1.1310223

66. Chen, J., Zhang, G., Li, B.: Molecular dynamics simulations of heat conduction in nanostructures: effect of heat bath. J. Phys. Soc. Jpn. **79**(7), 074604 (2010). https://doi.org/10.1143/JPSJ.79.074604

67. Alaghemandi, M., Müller-Plathe, F., Böhm, M.C.: Thermal conductivity of carbon nanotube-polyamide-6,6 nanocomposites: reverse non-equilibrium molecular dynamics simulations. J. Chem. Phys. **135**(18), 11B606 (2011). https://doi.org/10.1063/1.3660348

68. Müller-Plathe, F.: A simple nonequilibrium molecular dynamics method for calculating the thermal conductivity. J. Chem. Phys. **106**(14), 6082–6085 (1997). https://doi.org/10.1063/1.473271

69. Turney, J.E., McGaughey, A.J.H., Amon, C.H.: Assessing the applicability of quantum corrections to classical thermal conductivity predictions. Phys. Rev. B **79**(22), 224305 (2009). https://doi.org/10.1103/PhysRevB.79.224305

70. Lü, J.T., Wang, J.S.: Coupled electron-phonon transport from molecular dynamics with quantum baths. J. Phys.: Condens. Matter **21**, 025503 (2009). https://doi.org/10.1088/0953-8984/21/2/025503

71. Pecchia, A., Penazzi, G., Salvucci, L., Di Carlo, A.: Non-equilibrium Green's functions in density functional tight binding: method and applications. New J. Phys. **10**(6), 065022 (2008). https://doi.org/10.1088/1367-2630/10/6/065022

72. Zeng, J., Zhang, R.Q., Treutlein, H. (eds.): Quantum simulations of materials and biological systems. Springer, Dordrecht (2012)

73. Zhou, H.B., Cai, Y.Q., Zhang, G., Zhang, Y.W.: Quantum thermal transport in stanene. Phys. Rev. B **94**, 045423 (2016). https://doi.org/10.1103/PhysRevB.94.045423

74. Wang, J.S., Agarwalla, B.K., Li, H., Thingna, J.: Nonequilibrium Green's function method for quantum thermal transport. Front. Phys. **9**(6), 673–697 (2014). https://doi.org/10.1007/s11467-013-0340-x

75. Hong, Y., Zhang, J., Zeng, X.C.: Thermal transport in phosphorene and phosphorene-based materials: a review on numerical studies. Chin. Phys. B **27**, 036501 (2018). https://doi.org/10.1088/1674-1056/27/3/036501

76. Caroli, C., Combescot, R., Nozieres, P., Saint-James, D.: Direct calculation of the tunneling current. J. Phys. C: Solid State Phys. **4**(8), 916 (1971). https://doi.org/10.1088/0022-3719/4/8/018
77. Wang, J.S., Wang, J., Lü, J.T.: Quantum thermal transport in nanostructures. Eur. Phys. J. B **62**(4), 381–404 (2008). https://doi.org/10.1140/epjb/e2008-00195-8
78. Markussen, T., Jauho, A.P., Brandbyge, M.: Heat conductance is strongly anisotropic for pristine silicon nanowires. Nano Lett. **8**(11), 3771–3775 (2008). https://doi.org/10.1021/nl8020889
79. Luisier, M.: Investigation of thermal transport degradation in rough Si nanowires. J. Appl. Phys. **110**(7), 074510 (2011). https://doi.org/10.1063/1.3644993
80. Yamamoto, K., Ishii, H., Kobayashi, N., Hirose, K.: Thermal conductance calculations of silicon nanowires: comparison with diamond nanowires. Nanoscale Res. Lett. **8**(1), 256 (2013). https://doi.org/10.1186/1556-276X-8-256
81. Royo, M., Rurali, R.: Tuning thermal transport in Si nanowires by isotope engineering. Phys. Chem. Chem. Phys. **18**(37), 26262–26267 (2016). https://doi.org/10.1039/C6CP04581B
82. Bulusu, A., Walker, D.G.: Quantum modeling of thermoelectric performance of strained Si/Ge/Si superlattices using the nonequilibrium Green's function method. J. Appl. Phys. **102**(7), 073713 (2007). https://doi.org/10.1063/1.2787162
83. Xu, Y., Chen, X., Gu, B.L., Duan, W.: Intrinsic anisotropy of thermal conductance in graphene nanoribbons. Appl. Phys. Lett. **95**(23), 233116 (2009). https://doi.org/10.1063/1.3272678
84. Bachmann, M., Czerner, M., Edalati-Boostan, S., Heiliger, C.: Ab initio calculations of phonon transport in ZnO and ZnS. Eur. Phys. J. B **85**(5), 146 (2012). https://doi.org/10.1140/epjb/e2012-20503-y
85. Lü, J.T., Zhou, H., Jiang, J.W., Wang, J.S.: Effects of electron-phonon interaction on thermal and electrical transport through molecular nano-conductors. AIP Adv. **5**(5), 053204 (2015). https://doi.org/10.1063/1.4917017
86. Li, X., Yang, R.: Effect of lattice mismatch on phonon transmission and interface thermal conductance across dissimilar material interfaces. Phys. Rev. B **86**(5), 054305 (2012). https://doi.org/10.1103/PhysRevB.86.054305
87. Shi, L., Li, D., Yu, C., Jang, W., Kim, D., Yao, Z., Kim, P., Majumdar, A.: Measuring thermal and thermoelectric properties of one-dimensional nanostructures using a microfabricated device. J. Heat Transfer **125**(5), 881–888 (2003). https://doi.org/10.1115/1.1597619
88. Balandin, A.A., Ghosh, S., Bao, W.Z., Calizo, I., Teweldebrhan, D., Miao, F., Lau, C.N.: Superior thermal conductivity of single-layer graphene. Nano Lett. **8**(3), 902–907 (2008). https://doi.org/10.1021/nl0731872
89. Lu, L., Yi, W., Zhang, D.L.: 3ω method for specific heat and thermal conductivity measurements. Rev. Sci. Instrum. **72**, 2996 (2001). https://doi.org/10.1063/1.1378340
90. Cahill, D.G.: Analysis of heat flow in layered structures for time-domain thermoreflectance. Rev. Sci. Instrum. **75**, 5119 (2004). https://doi.org/10.1063/1.1819431
91. Kim, P., Shi, L., Majumdar, A., McEuen, P.L.: Thermal transport measurements of individual multiwalled nanotubes. Phys. Rev. Lett. **87**, 215502 (2001). https://doi.org/10.1103/PhysRevLett.87.215502
92. Pettes, M.T., Shi, L.: Thermal and structural characterizations of individual single-, double-, and multi-walled carbon nanotubes. Adv. Funct. Mater. **19**(24), 3918–3925 (2009). https://doi.org/10.1002/adfm.200900932
93. Li, D., Wu, Y., Kim, P., Shi, L., Yang, P., Majumdar, A.: Thermal conductivity of individual silicon nanowires. Appl. Phys. Lett. **83**, 2934 (2003). https://doi.org/10.1063/1.1616981
94. Wang, X., Yang, J., Xiong, Y., Huang, B., Xu, T.T., Li, D., Xu, D.: Measuring nanowire thermal conductivity at high temperatures. Meas. Sci. Technol. **29**, 025001 (2018). https://doi.org/10.1088/1361-6501/aa9389
95. Tinh Bui, C., Xie, R., Zheng, M., Zhang, Q., Sow, C.H., Li, B., Thong, J.T.: Diameter-dependent thermal transport in individual ZnO nanowires and its correlation with surface coating and defects. Small **8**(5), 738–745 (2012). https://doi.org/10.1002/smll.201102046

96. Pettes, M.T., Jo, I., Yao, Z., Shi, L.: Influence of polymeric residue on the thermal conductivity of suspended bilayer graphene. Nano Lett. **11**(3), 1195–1200 (2011). https://doi.org/10.102 1/nl104156y

97. Seol, J.H., Jo, I., Moore, A., Lindsay, L., Aitken, Z.H., Pettes, M.T., Li, X., Yao, Z., Huang, R., Broido, D.: Two-dimensional phonon transport in supported graphene. Science **328**(5975), 213–216 (2010). https://doi.org/10.1126/science.1184014

98. Hsin, C.L., Wingert, M., Huang, C.W., Guo, H., Shih, T.J., Suh, J., Wang, K., Wu, J., Wu, W.W., Chen, R.: Phase transformation and thermoelectric properties of bismuth-telluride nanowires. Nanoscale **5**(11), 4669–4672 (2013). https://doi.org/10.1039/C3NR00876B

99. Hochbaum, A.I., Chen, R., Delgado, R.D., Liang, W., Garnett, E.C., Najarian, M., Majumdar, A., Yang, P.: Enhanced thermoelectric performance of rough silicon nanowires. Nature **451**(7175), 163 (2008). https://doi.org/10.1038/nature06381

100. Lee, E.K., Yin, L., Lee, Y., Lee, J.W., Lee, S.J., Lee, J., Cha, S.N., Whang, D., Hwang, G.S., Hippalgaonkar, K., Majumdar, A., Yu, C., Choi, B.L., Kim, J.M., Kim, K.: Large thermoelectric figure-of-merits from SiGe nanowires by simultaneously measuring electrical and thermal transport properties. Nano Lett. **12**(6), 2918–2923 (2012). https://doi.org/10.1021/nl 300587u

101. Xu, X., Pereira, L.F.C., Wang, Y., Wu, J., Zhang, K., Zhao, X., Bae, S., Bui, C. T., Xie, R., Thong, J.T.L., Hong, B.H., Loh, K.P., Donadio, D., Li, B., Özyilmaz, B.: Length-dependent thermal conductivity in suspended single-layer graphene. Nat. Commun. **5**, 3689 (2014). https://doi.org/10.1038/ncomms4689

102. Jo, I., Pettes, M.T., Ou, E., Wu, W., Shi, L.: Basal-plane thermal conductivity of few-layer molybdenum disulfide. Appl. Phys. Lett. **104**, 201902 (2014). https://doi.org/10.1063/1.487 6965

103. Liu, D., Xie, R.G., Yang, N., Li, B.W., Thong, J.T.L.: Profiling nanowire thermal resistance with a spatial resolution of nanometers. Nano Lett. **14**, 806–812 (2014). https://doi.org/10.1 021/nl4041516

104. Cai, W., Moore, A.L., Zhu, Y., Li, X., Chen, S., Shi, L., Ruoff, R.S.: Thermal transport in suspended and supported monolayer graphene grown by chemical vapor deposition. Nano Lett. **10**(5), 1645–1651 (2010). https://doi.org/10.1021/nl9041966

105. Li, Q.Y., Xia, K.L., Zhang, J., Zhang, Y.Y., Li, Q.Y., Takahashi, K., Zhang, X.: Measurement of specific heat and thermal conductivity of supported and suspended graphene by a comprehensive Raman optothermal method. Nanoscale **9**(30), 10784–10793 (2017). https://doi.org/ 10.1039/C7NR01695F

106. Chen, S., Moore, A.L., Cai, W., Suk, J.W., An, J., Mishra, C., Amos, C., Magnuson, C.W., Kang, J., Shi, L., Ruoff, R.S.: Raman measurements of thermal transport in suspended monolayer graphene of variable sizes in vacuum and gaseous environments. ACS Nano **5**(1), 321–328 (2010). https://doi.org/10.1021/nn102915x

107. Yan, R., Simpson, J.R., Bertolazzi, S., Brivio, J., Watson, M., Wu, X., Kis, A., Luo, T., Hight Walker, A.R., Xing, H.G.: Thermal conductivity of monolayer molybdenum disulfide obtained from temperature-dependent Raman spectroscopy. ACS Nano **8**(1), 986–993 (2014). https:// doi.org/10.1021/nn405826k

108. Poborchii, V., Uchida, N., Miyazaki, Y., Tada, T., Geshev, P.I., Utegulov, Z.N., Volkov, A.: A simple efficient method of nanofilm-on-bulk-substrate thermal conductivity measurement using Raman thermometry. Int. J. Heat Mass Transf. **123**, 137–142 (2018). https://doi.org/1 0.1016/j.ijheatmasstransfer.2018.02.074

109. Doerk, G.S., Carraro, C., Maboudian, R.: Single nanowire thermal conductivity measurements by Raman thermography. ACS Nano **4**(8), 4908–4914 (2010). https://doi.org/10.1021/nn101 2429

110. Hsu, I.K., Kumar, R., Bushmaker, A., Cronin, S.B., Pettes, M.T., Shi, L., Brintlinger, T., Fuhrer, M.S., Cumings, J.: Optical measurement of thermal transport in suspended carbon nanotubes. Appl. Phys. Lett. **92**(6), 063119 (2008). https://doi.org/10.1063/1.2829864

111. Yu, D., Li, S., Qi, W., Wang, M.: Temperature-dependent Raman spectra and thermal conductivity of multi-walled MoS_2 nanotubes. Appl. Phys. Lett. **111**(12), 123102 (2017). https://do i.org/10.1063/1.5003111

112. Beechem, T., Yates, L., Graham, S.: Error and uncertainty in Raman thermal conductivity measurements. Rev. Sci. Instrum. **86**(4), 041101 (2015). https://doi.org/10.1063/1.4918623

113. Cahill, D.G.: Thermal conductivity measurement from 30 to 750 K: the 3ω method. Rev. Sci. Instrum. **61**, 802 (1990). https://doi.org/10.1063/1.1141498

114. Lee, K.M., Choi, T.Y., Lee, S.K., Poulikakos, D.: Focused ion beam-assisted manipulation of single and double β-SiC nanowires and their thermal conductivity measurements by the four-point-probe 3ω method. Nanotechnology **21**(12), 125301 (2010). https://doi.org/10.108 8/0957-4484/21/12/125301

115. Rojo, M.M., Calero, O.C., Lopeandia, A.F., Rodriguez-Viejo, J., Martín-Gonzalez, M.: Review on measurement techniques of transport properties of nanowires. Nanoscale **5**(23), 11526–11544 (2013). https://doi.org/10.1039/C3NR03242F

116. Pennelli, G., Dimaggio, E., Macucci, M.: Improvement of the 3ω thermal conductivity measurement technique for its application at the nanoscale. Rev. Sci. Instrum. **89**(1), 016104 (2018). https://doi.org/10.1063/1.5008807

117. Mallet, J., Molinari, M., Martineau, F., Delavoie, F., Fricoteaux, P., Troyon, M.: Growth of silicon nanowires of controlled diameters by electrodeposition in ionic liquid at room temperature. Nano Lett. **8**(10), 3468–3474 (2008). https://doi.org/10.1021/nl802352e

118. Zhang, R.Q., Lifshitz, Y., Lee, S.T.: Oxid-assisted growth of semiconducting nanowires. Adv. Mater. **15**(7–8), 635–640 (2003). https://doi.org/10.1002/adma.200301641

119. Novoselov, K.S., Geim, A.K., Morozov, S.V., Jiang, D., Zhang, Y., Dubonos, S.V., Grigorieva, I.V., Firsov, A.A.: Electric field effect in atomically thin carbon films. Science **306**(5696), 666–669 (2004). https://doi.org/10.1126/science.1102896

120. Zhang, Y., Zhang, L., Zhou, C.: Review of chemical vapor deposition of graphene and related applications. Acc. Chem. Res. **46**(10), 2329–2339 (2013). https://doi.org/10.1021/ar300203n

Chapter 3
Thermal Stability and Phonon Thermal Transport in Spherical Silicon Nanoclusters

3.1 Structure and Thermal Stability of Pristine Silicon Nanoclusters

Silicon nanoparticles or nanoclusters have received wide attention in the scientific community because of their special physical properties and promising applications in opto-electronic devices [1, 2], photovoltaic solar cells [3], and thermoelectric devices [4]. The structural and thermodynamic properties of nanoclusters are fundamentally important for optimizing the preparation process [5, 6], because the performance and stability of nanomaterials under different temperatures are closely related to their thermodynamic properties. Determining the melting behaviour along the crystal-liquid phase equilibrium line on the phase diagram is crucial, along with understanding the microscopic mechanisms underlying the transition between solid and liquid phases at high temperature [7]. As the size decreases to the nanometer scale, surfaces become a large fraction of the materials; moreover, they can strongly affect the overall properties including the thermodynamic properties of a material, which are dramatically different from the bulk phase [8]. Theoretical research on silicon nanoparticles has been dedicated heavily to the size-dependent electronic and optical properties [9–19]. To our knowledge, very few studies have been conducted on the melting and thermal stability of the silicon nanocrystals.

3.1.1 Structures and Melting Properties

Experimental [20, 21] and theoretical [22] studies have demonstrated the attenuation of the melting temperature of nanoparticles below that of the bulk materials in many metals. However, a recent investigation has reported an interesting observation on the

© The Author(s), under exclusive licence to Springer Nature Singapore Pte Ltd. 2018   41
H.-P. Li and R.-Q. Zhang, *Phonon Thermal Transport in Silicon-Based Nanomaterials*,
SpringerBriefs in Physics, https://doi.org/10.1007/978-981-13-2637-0_3

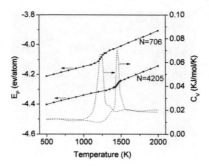

Fig. 3.1 Temperature dependence of potential energy per atom (E_P) and heat capacity (C_V) for Si_{706} and Si_{4205}, simulated by Tersoff-ARK potential. Reprinted with permission from Ref. [26], copyright (2017) by Springer Nature

melting point of tin clusters above that of the bulk material [23]. These size-dependent melting properties of metallic clusters are related to the close relationship between the melting and the structure features [20, 24, 25]. However, the melting process and its microscopic mechanism have not been carefully examined for medium-sized silicon nanoclusters. Experimental investigation on the melting process of different-sized nanoclusters/nanoparticles is extremely difficult, but molecular dynamics (MD) simulations can provide valuable insights into the structure change and energy of the nanocluster during heating. Li et al. [26] studied the structural and melting properties of several silicon spherical nanoclusters containing a maximum of 13,407 atoms using empirical MD simulations with Tersoff-ARK potential.

Figure 3.1 shows the temperature dependence of potential energy per atom and heat capacity for silicon spherical clusters with 706 atoms and with 4205 atoms, respectively. A dynamic phase coexistence in melting is found near the jump at which the number of liquid-like atoms increases with rising temperature. Coexistence melting will shift to a higher temperature as the size of the cluster increases. For example, the melting point of Si_{4205} is about 1450 K. When the number of atoms is decreased to 706, the melting point reaches about 1240 K. Moreover, a sharp peak of the heat capacity clearly shows a first-order behaviour of the melting in nanoclusters. However, the melting of crystalline nanoparticles proceeds over a temperature range due to surface effect, indicating that the melting of nanoclusters is heterogeneous [5], unlike the homogeneous melting of bulk silicon.

Li et al. [26] also studied the melting mechanism of silicon nanoclusters. Figure 3.2 depicts snapshots of the Si_{706} cluster at certain temperatures, together with the corresponding radial distribution functions (RDF). These snapshots qualitatively provide clear evidence of the increasing liquid skin thickness on the cluster surface from 500 to 1300 K and demonstrates that the inner region remains orderly until the transition temperature of approximately 1240 K. This outcome provides proof of the surface melting of silicon nanoclusters. Further confirmation of such surface melting is also obtained from the RDF of the melting process. At 500 K, a typical crystal structure pattern occurs. Before melting (up to 1200 K), the atom types of first-nearest neigh-

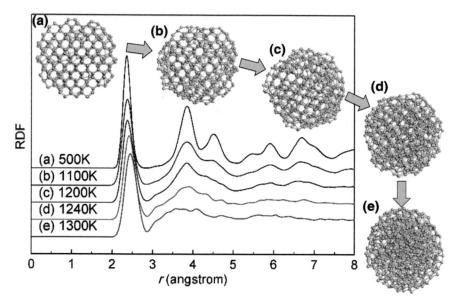

Fig. 3.2 Snapshots and RDF of Si$_{706}$ cluster at different temperatures. Readapted with permission from Ref. [26], copyright (2017) by Springer Nature

bours are unchanged. However, the peaks become less intensive and broader because of the increasing heat motion. After melting (at 1300 K), the second and third peaks disappear in the RDF, indicating that the short-range order of atoms changed significantly. Hence, a notable short-range order change may occur between 1200 and 1300 K. The changes in the RDF at melting are related to the changes of the structural order. With the increase of amplitudes of atom motions, atoms can overcome structural constraints and then melt into a liquid state.

Li et al. [26] examined the size dependence of the melting point for pure silicon spherical clusters, as shown in Fig. 3.3. The melting temperature shows a linear function of $N^{-1/3}$, where N is the number of atoms. Extrapolating the results for infinite size ($N \rightarrow \infty$) yields a predicted value of 1721 K for bulk silicon with a free surface, which is much closer to the experimental value of 1687 K than the Tersoff-ARK predicted value (2150 K) and the Stillinger-Weber potential predicted value (2370 K) for bulk silicon without a surface. This discrepancy shows the significance of the surface effect in the melting of finite clusters.

3.1.2 Size-Induced Structural Transition

Nanoscale clusters (also called nanoparticles) with diameters of 1–10 nm have attracted considerable scientific interest because they can emerge as a bridge between bulk materials and atomic/molecular structures. The silicon nanocluster structure can

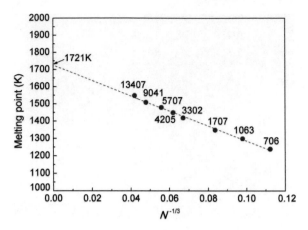

Fig. 3.3 Size dependence of the melting point in silicon spherical clusters. The solid circle indicates the melting point of different sized clusters. The numbers near the solid circles denote the numbers of cluster atoms. The dashed line shows the best fit to a linear function of $N^{-1/3}$. Reprinted with permission from Ref. [26], copyright (2017) by Springer Nature

be expected to change gradually from the bulk diamond structure to a non-diamond structure with the decrease in cluster size, but determining the accurate structural transition size is difficult [8]. Unsaturated silicon nanoclusters are naturally unstable and can easily develop into structures of various possible morphologies with coordination largely deviating from four of the bulk materials.

By using simulated annealing MD simulations, Li and Zhang [27] studied the structural transition characteristics in the nanosized silicon clusters. As shown in Fig. 3.4, when the cluster diameters increase from 1.80 to 3.28 nm, the structures of the clusters change from an amorphous state to a crystalline core/amorphous shell state and then to a crystalline state. The smallest crystalline silicon particle diameter ranges from 2.40 to 2.60 nm approximately, which is in good agreement with the reported experimental and theoretical results. Mélinon et al. [28] found that the critical size is above 2.00 nm (~200 atoms) through the transmission electron microscopy. However, Hofmeister et al. [29] verified that the smallest crystalline region in the annealed amorphous silicon powders is 2.50 nm. A subsequent photoluminescence (PL) study of silicon nanocrystals revealed that the smallest nanocrystal in which PL follows the quantum confinement model is approximately 2.80 nm in diameter [30]. Yu et al. [8] also theoretically calculated the cohesive energy of silicon nanoclusters with different sizes after tight-binding MD annealing and predicted that the transition to the diamond structure is estimated to occur in the diameter range from 2.30 to 2.70 nm via extrapolation of the cohesive energy curve.

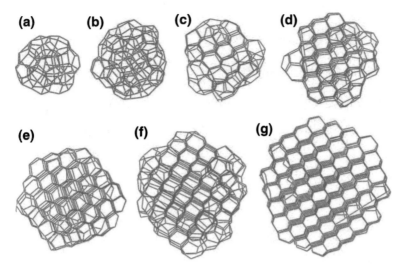

Fig. 3.4 Geometries of the annealed silicon clusters with diameters: **a** $r = 1.80$ nm, **b** $r = 2.12$ nm, **c** $r = 2.28$ nm, **d** $r = 2.42$ nm, **e** $r = 2.63$ nm, **f** $r = 2.88$ nm, and **g** $r = 3.28$ nm. Adapted from Ref. [27]

3.1.3 Structural Stability from Hydrogenation

Hydrogenation can stabilize silicon nanoparticles/nanoclusters surfaces against oxidation [31], after simply etching the silicon particles surfaces in aqueous hydrofluoric acid followed by rinsing in water [32]. Hydrogen (H)-terminated silicon surfaces are constantly attracting systems for intensive study given their great technological importance in microelectronics. The theoretical study by Zhang et al. [33] showed that the surface hydrogenation effects of silicon small particles are important for their electronic structures.

Yu et al. [34] examined the structures of hydrogenated silicon nanoclusters using tight-binding MD simulations. They found that the structural properties of the H saturated silicon nanocrystals have less size effect than their electronic properties. Surface relaxation is small in the fully H saturated silicon nanocrystals. Only atoms in the outermost two or three layers exhibit small lattice contractions of 0.01–0.02 Å. Surface relaxation mainly depends on the local environment. Taking the small $Si_{100}H_x$ nanoclusters for instance, simulated annealing simulations have been performed to obtain energetically favourable structures, as shown in Fig. 3.5. Fully hydrogenated clusters are proven to be the most stable structures compared to those that are partially hydrogenated. Removing up to 50% of the total H atoms only causes lattice distortions to the crystal structure while retaining the tetrahedral structures. Conversely, removing more than 70% of the total H atoms allows the clusters to evolve into more compact structures.

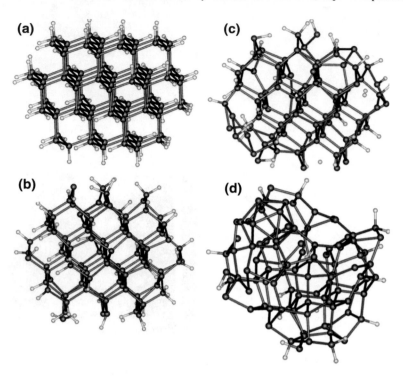

Fig. 3.5 Optimized structures of $Si_{100}H_x$ clusters from simulated annealing. **a** Fully H saturated $Si_{100}H_{86}$, **b** $Si_{100}H_{60}$, **c** $Si_{100}H_{40}$, and **d** $Si_{100}H_{20}$. Reprinted with permission from Ref. [34], copyright (2002) by AIP Publishing

3.2 Phonon Thermal Conductivity of Pristine Silicon Nanoclusters

As discussed in the last section, due to the size effect, silicon nanoclusters offer unique structural and thermal properties which are important for future applications in energy conversion. Recently, an increasing number of studies [35–37] have shown that silicon nanostructures, including nanoparticles, can be promising materials for thermoelectric applications because of their reduced thermal conductivity. Knowledge of thermal transport properties of silicon nanoclusters is essential for optimizing the operation of their nanodevices. However, compared with the research on their electronic and optical properties, that on the thermal conductivity of semiconductor nanoclusters/nanoparticles requires much more effort.

Li and Zhang [27] have recently investigated the phonon thermal transport properties of spherical silicon nanoclusters using Tersoff potential based equilibrium MD simulations. Figure 3.6 shows the time dependence (0–300 ps) of the normalized heat current autocorrelation functions (NHCACF) at 300 K for bulk crystalline silicon and the Si_{465} cluster, respectively. For bulk silicon, the NHCACF decays very

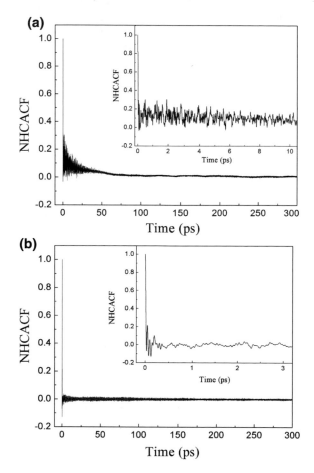

Fig. 3.6 Typical curves for NHCACF of **a** bulk silicon and **b** Si$_{465}$ cluster at 300 K. The insets show the corresponding NHCACF curves at the small timescale. Adapted from Ref. [27]

rapidly in the beginning (~0.20 ps), followed by a much slower decay. However, the NHCACF curve for the nanocluster decays to zero at 0.50 ps. The rapid decay corresponds to the reduced contribution of short-wavelength phonons to the phonon thermal conductivity, whereas the slower decay corresponds to the reduced contribution from long-wavelength phonons, which is the predominant contribution to the phonon thermal conductivity.

Figure 3.7 plots the calculated room temperature phonon thermal conductivities of spherical silicon nanoclusters with different diameters. The calculated phonon thermal conductivities of the silicon nanoclusters considered are three orders of magnitude lower than the thermal conductivity of the bulk (~150 W/mK [38]) at 300 K, which is consistent with recent experimental measurements [39, 40] (Fig. 3.7). In addition, a linear relationship is observed between the cluster diameter (d) and the

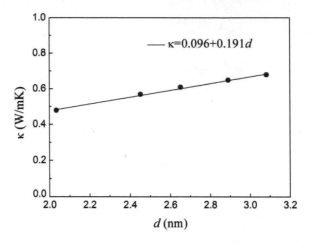

Fig. 3.7 Size dependence of phonon thermal conductivity of spherical silicon nanoclusters at 300 K. Adapted from Ref. [27]

thermal conductivity (κ) for the cluster considered in accordance with the Stillinger-Weber potential study [5]. The phonon mean free path (Λ) is the average distance a phonon travels between two collisions. According to kinetic theory of gas (see Chap. 2), the κ strongly depends on Λ in a perfect crystal material. In bulk silicon, Λ is ~260 nm at room temperature [41, and references therein]. Such a sizeable Λ results in the high thermal conductivity. However, in nanoclusters with diameters of several nanometers, the strong effect of phonon-boundary scattering remarkably reduces Λ. Here, d (<5 nm) is significantly smaller than Λ in bulk silicon at room temperature. When d is smaller than Λ, Λ will be limited by the cluster boundary. The phonon thermal conductivity can be written as $\kappa \propto C v_g d$, where C is average specific heat capacity and v_g is the average group velocity of phonons [42], which accounts for the linear relationship observed. This equation above is independent of the phonon frequency and determined only by the geometrical properties of the crystal particle [39]. In this case, the thermal conductivity becomes *ballistic* and thus shows a significant size-dependent effect.

Li [43] also investigated the temperature dependence of the phonon thermal conductivity of silicon nanocrystals (an example shown in Fig. 3.8). When the temperature is increased from 50 to 1000 K, the phonon thermal conductivity of the Si$_{465}$ cluster decreases sharply at low temperatures (<300 K) and decreases more slowly over 300 K, indicating that a stronger phonon-phonon scattering dominantly reduces the Λ of phonons at higher temperatures [44].

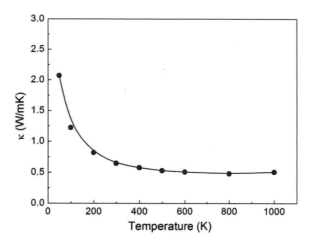

Fig. 3.8 Temperature dependence of the phonon thermal conductivity (κ) of Si$_{465}$ cluster. Adapted from Ref. [43]

References

1. Ito, M., Imakita, K., Fujii, M., Hayashi, S.: Nonlinear optical properties of phosphorus-doped silicon nanocrystals/nanoclusters. J. Phys. D Appl. Phys. **43**, 505101 (2010). https://doi.org/1 0.1088/0022-3727/43/50/505101
2. Wang, M., Li, D., Yuan, Z., Yang, D., Que, D.: Photoluminescence of Si-rich silicon nitride: defect-related states and silicon nanoclusters. Appl. Phys. Lett. **90**, 131903 (2007). https://do i.org/10.1063/1.2717014
3. Meinardi, F., Ehrenberg, S., Dhamo, L., Carulli, F., Mauri, M., Bruni, F., Brovelli, S.: Highly efficient luminescent solar concentrators based on earth-abundant indirect-bandgap silicon quantum dots. Nat. Photon. **11**(3), 177–185 (2017). https://doi.org/10.1038/nphoton.2017.5
4. Joshi, G., Lee, H., Lan, Y., Wang, X., Zhu, G., Wang, D., Gould, R.W., Cuff, D.C., Tang, M.Y., Dresselhaus, M.S., Chen, G., Ren, Z.: Enhanced thermoelectric figure-of-merit in nanostructured p-type silicon germanium bulk alloys. Nano Lett. **8**(12), 4670–4674 (2008). https://doi.org/10.1021/nl8026795
5. Fang, K.C., Weng, C.I.: An investigation into the melting of silicon nanoclusters using molecular dynamics simulations. Nanotechnology **16**(2), 250–256 (2005). https://doi.org/10.1088/0 957-4484/16/2/012
6. Sang, L.V., Hoang, V.V., Tranh, D.T.N.: Melting of crystalline Si nanoparticle investigated by simulation. Eur. Phys. J. D **69**, 208 (2015). https://doi.org/10.1140/epjd/e2015-60153-1
7. Dozhdikov, V.S., Basharin, A.Y., Levashov, P.R.: Two-phase simulation of the crystalline silicon melting line at pressures from −1 to 3 GPa. J. Chem. Phys. **137**(5), 054502 (2012). https://do i.org/10.1063/1.4739085
8. Yu, D.K., Zhang, R.Q., Lee, S.T.: Structural transition in nanosized silicon clusters. Phys. Rev. B **65**, 245417 (2002). https://doi.org/10.1103/PhysRevB.65.245417
9. Zhang, R.Q.: Growth mechanisms and novel properties of silicon nanostructures from quantum-mechanical calculations. Springer, Berlin Heidelberg (2014)
10. Zeng, J., Zhang, R.Q., Treutlein, H. (eds.): Quantum simulations of materials and biological systems. Springer, Dordrecht (2012)
11. Wu, Q., Wang, X.H., Niehaus, T.A., Zhang, R.Q.: Boundary and symmetry determined exciton distribution in two dimensional silicon nanosheets. J. Phys. Chem. C **118**(35), 20070–20076 (2014). https://doi.org/10.1021/jp501433t

12. Li, Q.S., Zhang, R.Q., Lee, S.T.: Stabilizing excited-state silicon nanoparticle by surface oxidation. Appl. Phys. Lett. **91**, 043106 (2007). https://doi.org/10.1063/1.2762296
13. Li, Q.S., Zhang, R.Q., Niehaus, T.A., Frauenheim, Th, Lee, S.T.: Theoretical studies on optical and electronic properties of propionic-acid-terminated silicon quantum dots. J. Chem. Theory Comput. **3**(4), 1518–1526 (2007). https://doi.org/10.1021/ct700041v
14. Lu, A.J., Zhang, R.Q., Lee, S.T.: Tunable electronic band structures of hydrogen-terminated <1112> silicon nanowires. Appl. Phys. Lett. **92**, 203109 (2008). https://doi.org/10.1063/1.293 6088
15. Wang, X., Zhang, R.Q., Niehaus, T.A., Frauenheim, Th., Lee, S.T.: Hydrogenated silicon nanoparticles relaxed in excited states. J. Phys. Chem. C. **111**(34), 12588–12593 (2007). https://doi.org/110.1021/jp071384j
16. Yang, X.B., Zhang, R.Q.: Indirect-to-direct band gap transitions in phosphorus adsorbed <112> silicon nanowires. Appl. Phys. Lett. **93**, 173108 (2008). https://doi.org/10.1063/1.3012372
17. Li, H., Xu, H., Shen, X., Han, K., Bi, Z., Xu, R.: Size-, electric-field-, and frequency-dependent third-order nonlinear optical properties of hydrogenated silicon nanoclusters. Sci. Rep. **26**, 28067 (2016). https://doi.org/10.1038/srep28067
18. Imakita, K., Ito, M., Naruiwa, R., Fujii, M., Hayashi, S.: Ultrafast third order nonlinear optical response of donor and acceptor codoped and compensated silicon quantum dots. Appl. Phys. Lett. **101**, 041112 (2012). https://doi.org/10.1063/1.4739237
19. Ito, M., Imakita, K., Fujii, M., Hayashi, S.: Nonlinear optical properties of silicon nanoclusters/nanocrystals doped SiO_2 films: annealing temperature dependence. J. Appl. Phys. **108**, 063512 (2010). https://doi.org/10.1063/1.3480821
20. Dick, K., Dhanasekaran, T., Zhang, Z., Meisel, D.: Size-dependent melting of silica-encapsulated gold nanoparticles. J. Am. Chem. Soc. **124**(10), 2312–2317 (2002). https://doi.org/10.1021/ja017281a
21. Aguado, A., Jarrold, M.F.: Melting and freezing of metal clusters. Annu. Rev. Phys. Chem. **62**, 151–172 (2011). https://doi.org/10.1146/annurev-physchem-032210-103454
22. Alavi, S., Thompson, D.L.: Molecular dynamics simulations of the melting of aluminum nanoparticles. J. Phys. Chem. A **110**(4), 1518–1523 (2006). https://doi.org/10.1021/jp053318s
23. Shvartsburg, A.A., Jarrold, M.F.: Solid clusters above the bulk melting point. Phys. Rev. Lett. **85**(12), 2530–2532 (2000). https://doi.org/110.1103/PhysRevLett.85.2530
24. Samsonov, V.M., Sdobnyakov, N.Y., Vasilyev, S.A., Sokolov, D.N.: On the size dependence of the heats of melting of metal nanoclusters. Bull. Russ. Acad. Sci.: Phys. **80**(5), 494–496 (2016). https://doi.org/10.3103/S1062873816050166
25. Wang, J.L., Chen, X.S., Wang, G.H., Wang, B.L., Lu, W., Zhao, J.J.: Melting behavior in ultrathin metallic nanowires. Phys. Rev. B **66**, 085408 (2002). https://doi.org/10.1103/PhysRe vB.66.085408
26. Li, H.P., Xu, R.F., Bi, Z.T., Shen, X.P., Han, K.: Melting properties of medium-sized silicon nanoclusters: a molecular dynamics study. J. Electron. Mater. **46**(7), 3826–3830 (2017). https://doi.org/10.1007/s11664-016-5070-8
27. Li, H.P., Zhang, R.Q.: Size-dependent structural characteristics and phonon thermal transport in silicon nanoclusters. AIP Adv. **3**, 082114 (2013). https://doi.org/10.1063/1.4818591
28. Mélinon, P., Kéghélian, P., Prével, B., Perez, A., Guiraud, G., LeBrusq, J., Lermé, J., Pellarin, M., Broyer, M.: Nanostructured silicon films obtained by neutral cluster depositions. J. Chem. Phys. **107**, 10278 (1997). https://doi.org/10.1063/1.474168
29. Hofmeister, H., Dutta, J., Hofmann, H.: Atomic structure of amorphous nanosized silicon powders upon thermal treatment. Phys. Rev. B **54**, 2856 (1996). https://doi.org/10.1103/Phys RevB.54.2856
30. Ledoux, G., Guillois, O., Porterat, D., Reynaud, C., Huisken, F., Kohn, B., Paillard, V.: Photoluminescence properties of silicon nanocrystals as a function of their size. Phys. Rev. B **62**, 15942 (2000). https://doi.org/10.1103/PhysRevB.62.15942
31. Chabal, Y.J.: Hydride formation on the Si(100):H_2O surface. Phys. Rev. B **29**, 3677 (1984). https://doi.org/10.1103/PhysRevB.29.3677

32. Chabal, Y.J., Higashi, G.S., Raghavachari, K., Burrows, V.A.: Infrared spectroscopy of Si(111) and Si(100) surfaces after HF treatment: hydrogen termination and surface morphology. J. Vac. Sci. Technol. A **7**, 2104 (1989). https://doi.org/10.1116/1.575980
33. Zhang, R.Q., Costa, J., Bertran, E.: Role of structural saturation and geometry in the luminescence of silicon-based nanostructured materials. Phys. Rev. B **53**, 7847 (1996). https://doi.org/10.1103/PhysRevB.53.7847
34. Yu, D.K., Zhang, R.Q., Lee, S.T.: Structural properties of hydrogenated silicon nanocrystals and nanoclusters. J. Appl. Phys. **92**, 7453 (2002). https://doi.org/10.1063/1.1513878
35. Dresselhaus, M.S., Chen, G., Tang, M.Y., Yang, R.G., Lee, H., Wang, D.Z., Ren, Z.F., Fleurial, J.P., Gogna, P.: New directions for low–dimensional thermoelectric materials. Adv. Mater. **19**(8), 1043–1053 (2007). https://doi.org/10.1002/adma.200600527
36. Hochbaum, A.I., Chen, R., Delgado, R.D., Liang, W., Garnett, E.C., Najarian, M., Majumdar, A., Yang, P.: Enhanced thermoelectric performance of rough silicon nanowires. Nature **451**(7175), 163–167 (2008). https://doi.org/10.1038/nature06381
37. Petermann, N., Stötzel, J., Stein, N., Kessler, V., Wiggers, H., Theissmann, R., Schierning, G., Schmechel, R.: Thermoelectrics from silicon nanoparticles: the influence of native oxide. Eur. Phys. J. B **88**, 163 (2015). https://doi.org/10.1140/epjb/e2015-50594-7
38. Kremer, R.K., Graf, K., Cardona, M., Devyatykh, G.G., Gusev, A.V., Gibin, A.M., Inyushkin, A.V., Taldenkov, A.N., Pohl, H.J.: Thermal conductivity of isotopically enriched ^{28}Si: revisited. Solid State Commun. **131**(8), 499–503 (2004). https://doi.org/10.1016/j.ssc.2004.06.022
39. Juangsa, F.B., Muroya, Y., Ryu, M., Morikawa, J., Nozaki, T.: Thermal conductivity of silicon nanocrystals and polystyrene nanocomposite thin films. J. Phys. D Appl. Phys. **49**, 365303 (2016). https://doi.org/10.1088/0022-3727/49/36/365303
40. Wang, Z., Alaniz, J.E., Jang, W., Garay, J.E., Dames, C.: Thermal conductivity of nanocrystalline silicon: importance of grain size and frequency-dependent mean free paths. Nano Lett. **11**(6), 2206–2213 (2011). https://doi.org/10.1021/nl1045395
41. Li, H.P., De Sarkar, A., Zhang, R.Q.: Surface-nitrogenation-induced thermal conductivity attenuation in silicon nanowires. EPL **96**(5), 56007 (2011). https://doi.org/10.1209/0295-5075/96/56007
42. Balandin, A.A.: Nanoscale thermal management. IEEE Potentials **21**(1), 11–15 (2002). https://doi.org/10.1109/45.985321
43. Li, H.P.: Molecular dynamics simulations of phonon thermal transport in low-dimensional silicon structures. Doctoral dissertation, City University of Hong Kong (2012)
44. Zou, J., Balandin, A.: Phonon heat conduction in a semiconductor nanowire. J. Appl. Phys. **89**, 2932 (2001). https://doi.org/10.1063/1.1345515

Chapter 4
Phonon Thermal Transport in Silicon Nanowires and Its Surface Effects

With the development of nano-fabrication techniques, one-dimensional (1D) materials (such as nanowires [1, 2], nanotubes [3, 4], and quantum wires [5]) have been designed and synthesized over the past decades. Among them, silicon nanowires (SiNWs) have sparked growing interest from the scientific community and have wide applications in electronic, optoelectronic, and energy conversion devices [6–13] given their unique properties that differentiate them from bulk materials and their good compatibility with conventional silicon-based technology. In particular, early experimental studies on SiNWs have demonstrated their potential as efficient thermoelectric (TE) materials which bulk silicon fails to do [14, 15]. The thermoelectrics in nanowires have attracted wide attention because of their remarkably improved TE figure of merit compared to their bulk counterparts, which is mainly caused by the significant reduction of the thermal conductivity [16]. Hochbaum et al. [14] found that the figure of merit of SiNWs with a rough surface can approach 0.6 at 300 K arising from a remarkable 100-fold reduction in thermal conductivity over bulk values in rough SiNWs, while maintaining excellent electric properties. Such findings have inspired researchers to reconsider the applications of silicon in TE materials and provide a new strategy for increasing the TE figure of merit by decreasing thermal conductivity. Thus far, numerous approaches have been proposed to further reduce thermal conductivity in SiNWs for higher TE performance, such as the introduction of impurity scattering [17, 18], porous structure [19], surface roughness [20], strain engineering [21, 22], and surface functionalization previously proposed by the current authors [23, 24].

Understanding and controlling the thermal conductivity of nanomaterials play significant roles in promoting two fundamental applications [25]: thermal management and novel TE materials. Studies were conducted on the effects of size, cross-section, interface, defect, strain, and surface roughness on the thermal conductivity of SiNWs [20–33]. These studies propose the principles behind various strategies that decrease the thermal conductivity of SiNWs and provide guidance for the

© The Author(s), under exclusive licence to Springer Nature Singapore Pte Ltd. 2018
H.-P. Li and R.-Q. Zhang, *Phonon Thermal Transport in Silicon-Based Nanomaterials*,
SpringerBriefs in Physics, https://doi.org/10.1007/978-981-13-2637-0_4

experimental realization of these strategies. Readers interested in these topics can peruse recent review papers [7, 34–42]. Particularly, our more recent work [38] reviews the current advances in the research of phonon thermal transport in SiNWs, with focus on the surface effects of tunable phonon thermal conductivity.

4.1 Phonon Thermal Conductivity of Pristine SiNWs

The thermal conductivity of high-purity bulk silicon has been experimentally measured [43]. Silicon is infrequently applied for insulating hot objects as it conducts heat well, with a relatively high thermal conductivity of approximately 150 W/mK at room temperature. As mentioned in Chap. 2, the electronic contribution to thermal energy transport in silicon is negligible at room temperature. The phonon thermal conductivity of SiNWs can significantly differ from that of bulk Si because of the size effect and high surface-to-volume ratio of SiNWs. Early studies examined thermal transport in SiNWs through various theory-based methods, including equilibrium and non-equilibrium molecular dynamics (MD) simulations [44–48], Monte Carlo (MC) simulations [49, 50], the non-equilibrium Green's function formalism [51, 52], and the Boltzmann transport equation [53, 54]. Several works have demonstrated that the thermal conductivity of pristine nanowires is strongly dependent on their diameter, shape, and surface roughness. Volz and Chen [54] utilized equilibrium MD simulations to discover a remarkable reduction in the thermal conductivity of SiNWs. Two main factors are likely responsible for the low thermal conductivity of SiNWs. On the one hand, long-wavelength phonons that act as the main contributors to thermal conductivity are suppressed in the nanowire because the phonon mean free path is limited by the nanowire size. On the other hand, with the decrease in nanowire diameter, the surface-to-volume ratio increases, and phonon-boundary scattering thus becomes more dominant in the nanowire, further decreasing thermal conductivity.

Li et al. [55] presented the mesoscopic thermal transport measurements of individual SiNWs synthesized through the vapor-liquid-solid (VLS) method. As shown in Fig. 4.1a, the thermal conductivity of individual SiNWs decreases with a diminishing diameter. When the wire diameter decreases to below 150 nm, the thermal conductivity of the smooth VLS-grown SiNWs becomes significantly lower than the bulk value and can closely follow predictions based on Boltzmann transport theory, assuming that diffuse boundary scattering is the dominant phonon scattering mechanism. Thermal conductivity at 300 K ranges from 40 to 9 W/mK for SiNWs with diameters ranging from 115 to 22 nm, respectively. This finding provided the first experimental evidence for the importance of phonon boundary scattering in evaluating the thermal conductivity of NWs. The researchers also observed the anomalous temperature dependence of SiNWs with small diameters (Fig. 4.1b). To understand the physical mechanism that underlies the decreased thermal conductivity of SiNWs, different groups [30, 49, 56–60] used theoretical models and simulations to predict the phonon thermal conductivity of SiNWs. Mingo et al. [30] theoretically predicted

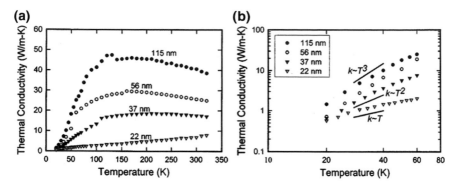

Fig. 4.1 **a** Measured thermal conductivity of SiNWs with different diameters. The number beside each curve denotes the corresponding wire diameter. **b** Low-temperature experimental data on a logarithmic scale. Reprinted with permission from Ref. [55], copyright (2003) by AIP Publishing

the temperature dependence of the thermal conductivity in SiNWs and found that the traditional Callaway and Holland approaches show substantial divergence with experimental data whereas the real dispersions approach yields good agreement with experiments for SiNWs between 37 and 115 nm wide.

4.2   Surface Effects of Phonon Thermal Conductivity of SiNWs

SiNWs exhibit a large surface-to-volume ratio, which increases with decreasing size. The electronic and thermal properties of SiNWs differ significantly from those of bulk silicon owing to the surface effect [61–64]. Consequently, surface effects and their characteristics have received extensive research attention given their key role in SiNWs. Surface atoms may play an important role in the overall thermal transport properties of SiNWs, particularly for thin SiNWs [23]. Therefore, the effect of surface treatments on phonon transport to further decrease the phonon thermal conductivity of SiNWs remains the focus of research interest. A parallel and simultaneous investigation of phonon transport is clearly needed given the extreme sensitivity of electron transport in nanostructures to the surface condition [65]. The thermal conductivity of SiNWs has been tuned through various theoretical and experimental surface engineering methodologies based on surface roughness, surface functionalization, surface/shell doping, surface disorder, and surface softening, as we summarized in [38].

4.2.1 Surface Roughness

As mentioned, Li et al. [55] observed that smooth VLS-grown SiNWs have low thermal conductivity. The thermal conductivity of electroless-etched (EE) SiNWs is five- to eight-fold lower than that of smooth VLS-grown SiNWs [14]. The surprising reduction in the thermal conductivity to a value below the Casimir limit cannot be explained by phonon-boundary scattering alone [66]. Hochbaum et al. [14] concluded that this unusually large reduction may be due to the surface roughness of the EE SiNWs. Subsequently, different groups analyzed the effect of surface roughness, although the exact mechanism of phonon roughness scattering remains unclear. Moore et al. [67] proposed a backscattering mechanism in SiNWs with sawtooth structures through MC simulations. Wang et al. [68] investigated the effect of surface roughness on thermal conductivity that accounts for the multiple scattering of phonons at a rough surface. Martin et al. [56, 69] examined the effect of NW surface roughness through perturbation theory and suggested that the thermal conductivity is limited by surface asperities. Recently, Malhotra et al. [57] have explored the impact of phonon surface scattering on the distribution of thermal energy across phonon wavelengths and mean free paths in Si and SiGe nanowires. They presented a rigorous and accurate description of surface phonon scattering and predicted heat transport in nanowires with different diameters and surface conditions. Xie et al. [58] utilized a kinetic model to investigate the anomalous thermal conductivity of SiNWs by focusing on the mechanism of phonon-boundary scattering.

Although current theoretical works have shed light on dependence of thermal conductivity on surface roughness, an experimental determination of the dependence of thermal conductivity on the surface roughness of SiNWs remains lacking. Kim et al. synthesized VLS-grown rough $Si_{1-x}Ge_x$ nanowires and measured their thermal conductivities [70] and found that the thermal conductivity of smooth $Si_{0.96}Ge_{0.04}$ nanowire is approximately four times higher than that of its rough counterpart. Theoretical analysis demonstrated that mid-wavelength phonons are scattered by rough surfaces. Lim et al. [66] experimentally studied the dependence of thermal conductivity on the surface roughness of SiNWs. Experimental observations proved that the surface of VLS-grown SiNWs is rough, with root-mean-square values that range from 0.3 to 5.0 nm. As shown in Fig. 4.2, thermal conductivity significantly decreases with increasing surface roughness. The coefficient of the surface roughness of single-phase crystalline SiNWs is well correlated with the reduction of the thermal conductivity.

4.2.2 Surface Functionalization

SiNWs are easily oxidized in air given the numerous dangling bonds distributed on their surfaces. Hydrogen termination on the surface is a natural consequence of the hydrofluoric acid treatment of synthesized SiNWs [63]. Park et al. [71] proposed

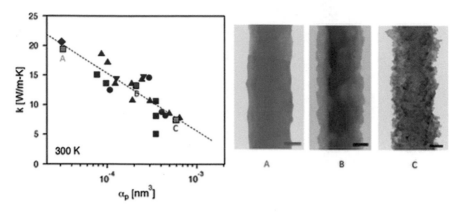

Fig. 4.2 Measured thermal conductivity as a function of roughness factor α_P at 300 K. Reprinted with permission from Ref. [66], copyright (2012) by the American Chemical Society

the selective surface functionalization of SiNWs through a nanoscale Joule-heating method. Surface chemical functionalization is crucial for electronic and thermal transports in SiNWs. Earlier theoretical studies showed that passivating atoms, such as hydrogen and nitrogen, on the surfaces of silicon nanostructures provide structurally stable structures and clean gap states [72]. The electrical and thermal conductivities of SiNW arrays and silicon membranes can be tuned through surface chemical modification [65, 73]. Our previous studies [23, 24] showed that the phonon thermal conductivity of hydrogenated SiNWs (H-SiNWs) is slightly higher than that of naked SiNWs (Fig. 4.3), but that of nitrogenated SiNWs is remarkably lower than that of fully hydrogenated SiNWs. As shown in Fig. 4.4, the functionalization of 14% of the surface silicon atoms by nitrogen atoms decreases the thermal conductivity of SiNWs by ~43%, and the functionalization of 29 and 43% of surface silicon atoms further decreases thermal conductivity by ~60% and ~75%, respectively. Markussen et al. [51] utilized atomic calculations for electron and phonon transport to investigate the properties of alkyl-functionalized SiNWs. Their results confirmed that phonon conductance significantly decreases relative to electronic conductance, thus providing SiNWs with a high thermoelectric figure of merit.

4.2.3 Surface/Shell Doping

Doping is an effective and feasible approach for decreasing thermal conductivity. Introducing dopant defects to the whole SiNWs can weaken electronic conductivity, but surface/shell doping can improve thermoelectric performance [74]. In contrast to traditional doping (in which dopant atoms are uniformly distributed inside nanowires), shell doping spatially confines dopant atoms within a few atomic layers in the shell region of a nanowire. Given their low thermal conductance and high

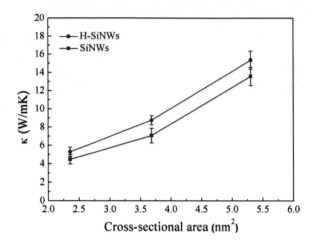

Fig. 4.3 Thermal conductivities of naked SiNWs and H-SiNWs at 300 K with different cross-section areas. Reprinted with permission from Ref. [23], copyright (2014) by EPLA

Fig. 4.4 Variation in the calculated longitudinal thermal conductivity of nitrogenated SiNWs ($\kappa_{\text{N-SiNW}}$) as a function of the surface nitrogenation ratio compared with that of naked SiNW (κ_{SiNW}). Reprinted with permission from Ref. [24], copyright (2011) by EPLA

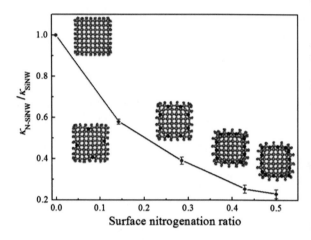

electrical conductance, shell-doped SiNWs containing high amounts of dopants have potential applications in the thermoelectric field [75].

The thermoelectric figure of merit of SiNWs can be increased by decreasing their thermal conductivity. Non-equilibrium MD simulations [17, 76] demonstrated that shell-doped SiNWs coated with germanium (Ge) have remarkably low thermal conductivity due to the impurity and interface scattering associated with their unique structure. Pan et al. [18] investigated surface Ge-doped SiNWs with diameters of approximately 100 nm. Figure 4.5 shows that the thermal conductivity of Ge-coated SiNW arrays decreased by 23% at room temperature relative to that of uncoated SiNWs. Subjecting the Ge-doped SiNW arrays to thermal annealing decreased the thermal conductivity by 44% because the surface-doped Ge interacts strongly with silicon, thus enhancing phonon scattering at the Si–Ge interface.

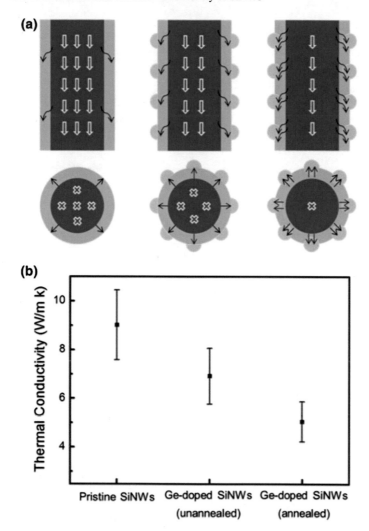

Fig. 4.5 a Schematic of the proposed phonon scattering mechanism in pristine SiNW, unannealed Ge-doped SiNW, and annealed Ge-doped SiNW. **b** Thermal conductivity of pristine SiNW, unannealed Ge-doped SiNW, and annealed Ge-doped SiNW. Reprinted with permission from Ref. [18], copyright (2015) by AIP Publishing

4.2.4 Surface Disorder

Apart from surface roughness, the surface structure of SiNWs can also be tailored by introducing surface disorder. An amorphous surface/shell layer can often form during the growth of NWs. Phonon scattering by surface roughness differs from that by an amorphous shell [37]. For example, under a very low temperature limit, the thermal conductivity of core-shell (CS) NWs with a thick amorphous surface

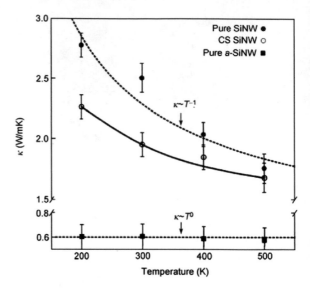

Fig. 4.6 Temperature dependence of the thermal conductivity (κ) of pure SiNW, CS SiNW, and a-SiNW. Reprinted with permission from Ref. [77], copyright (2014) by Springer Nature

shell is lower than that of rough crystalline NWs because of interface scattering and phonon-coherent resonance [76].

Liu et al. [77] performed early MD simulations to comparatively investigate the thermal conductivities of crystalline SiNWs and SiNWs with amorphous shells. As shown in Fig. 4.6, the thermal conductivity of amorphous SiNWs decreased by 80% relative to that of pure crystalline SiNW with the same size because of strong phonon scattering at the interface and the non-propagating diffusion of phonons in the amorphous region. The behavior of the thermal conductivity in crystalline SiNWs also displayed an unusual temperature dependence: the thermal conductivity of crystalline SiNW monotonically decreases with temperature following $1/T$. However, the thermal conductivity of amorphous SiNW remains nearly constant in the temperature range considered in the study and is analogous to that of its bulk counterpart. The temperature dependence of crystalline core-amorphous shell SiNW is considerably weaker than that of crystalline SiNW. These varied temperature trends reflect the different natures of the dominant scattering mechanisms in the materials. The thermal conductivity of crystalline SiNWs is proportional to $1/T$ because of the dominance of normal and Umklapp phonon scattering processes. By contrast, the thermal conductivity of amorphous SiNWs (a-SiNWs) is constant as structural disorder mainly contributes to scattering.

4.2.5   Surface Softening

Recently, the phonon softening effect has been discovered to play an important role in the thermal conductivity of solid solutions [78] and nanomaterials [79–82].

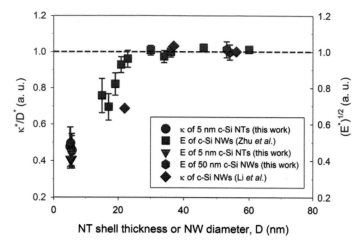

Fig. 4.7 Correlation between the thermal conductivity (κ) and elastic modulus (E) in crystalline nanotubes (NTs) and NWs. κ, D, and E have been normalized with their respective values for ~60 nm SiNWs and plotted as a function of D, where D is the shell thickness for NTs or the diameter for NWs. Normalized variables are labeled with asterisks. Reprinted with permission from Ref. [79], copyright (2015) by the American Chemical Society

For example, by solving the full Boltzmann transport equation, the significantly reduced phonon thermal conductivity was quantitatively examined in $Mg_2Si_{1-x}Sn_x$ because of the observed acoustic phonon softening [78]. Wingert et al. [79] experimentally confirmed that crystalline Si (c-Si) nanotubes (NTs) with shell thicknesses as thin as ~5 nm exhibit a low thermal conductivity of ~1.1 W/mK, which is lower than the apparent boundary scattering limit and is approximately 30% lower than the measured value for amorphous Si (a-Si) NTs with similar geometries (Fig. 4.7). Engineering the thermal transport properties of crystalline nanostructures beyond the phonon boundary scattering limit via the surface phonon softening effect is revealed to be possible due to the drastically reduced mechanical stiffness. This finding paves the way for new approaches to reduce nanostructure thermal conductivity.

References

1. Zhang, R.Q., Lifshitz, Y., Lee, S.T.: Oxide-assisted growth of semiconducting nanowires. Adv. Mater. **5**(7–8), 635–640 (2003). https://doi.org/10.1002/adma.200301641
2. Chang, P.C., Fan, Z.Y., Wang, D.W., Tseng, W.Y., Chiou, W.A., Hong, J., Lu, J.G.: ZnO nanowires synthesized by vapor trapping CVD method. Chem. Mater. **16**(24), 5133–5137 (2004). https://doi.org/10.1021/cm049182c
3. Iijima, S.: Helical microtubules of graphitic carbon. Nature **354**, 56–58 (1991). https://doi.org/10.1038/354056a0

4. Hata, K., Futaba, D.N., Mizuno, K., Namai, T., Yumura, M., Iijima, S.: Water-assisted highly efficient synthesis of impurity-free single-walled carbon nanotubes. Science 306(5700), 1362–1364 (2004). https://doi.org/10.1126/science.1104962
5. Ma, W.Q., Nötzel, R., Trampert, A., Ramsteiner, M., Zhu, H., Schönherr, H.P., Ploog, K.H.: Self-organized quantum wires formed by elongated dislocation-free islands in (In, Ga)As/GaAs(100). Appl. Phys. Lett. 78, 1297 (2001). https://doi.org/10.1063/1.1352047
6. Chen, L.J.: Silicon nanowires: the key building block for future electronic devices. J. Mater. Chem. 17(44), 4639–4643 (2007). https://doi.org/10.1039/B709983E
7. Rojo, M.M., Calero, O.C., Lopeandia, A.F., Rodriguez-Viejo, J., Martín-Gonzalez, M.: Review on measurement techniques of transport properties of nanowires. Nanoscale 5(23), 11526–11544 (2013). https://doi.org/10.1039/c3nr03242f
8. Weisse, J.M., Lee, C.H., Kim, D.R., Zheng, X.: Fabrication of flexible and vertical silicon nanowire electronics. Nano Lett. 12(6), 3339–3343 (2012). https://doi.org/10.1021/nl301659m
9. De, F.S., Trejo, A.B., Miranda, A., Salazar, F.P., Carvajal, E., Pérez, L.A., Cruz-Irisso, M.: Carbon monoxide sensing properties of B-Al- and Ga-doped Si nanowires. Nanotechnology 29(20), 204001 (2018). https://doi.org/10.1088/1361-6528/aab237
10. Yenchalwar, S.G., Rondiya, S.R., Shinde, P.N., Jadkar, S.R., Shelke, M.V.: Optical antenna effect on SiNWs/CuS photodiodes. Phys. Status Solidi A 214(5), 1600635 (2017). https://doi.org/10.1002/pssa.201600635
11. Gouda, A., Allam, N.K., Swillam, M.A.: Efficient fabrication methodology of wide angle black silicon for energy harvesting applications. RSC Adv. 7(43), 26974–26982 (2017). https://doi.org/10.1039/C7RA03568C
12. Bao, R.R., Zhang, C.Y., Zhang, X.J., Ou, X.M., Lee, C.S., Jie, J.S., Zhang, X.H.: Self-assembly and hierarchical patterning of aligned organic nanowire arrays by solvent evaporation on substrates with patterned wettability. ACS Appl. Mater. Interfaces. 5(12), 5757–5762 (2013). https://doi.org/10.1021/am4012885
13. Chandrasekaran, S., Nann, T., Voelcker, N.: Silicon nanowire photocathodes for photoelectrochemical hydrogen production. Nanomaterials 6(8), 144 (2016). https://doi.org/10.3390/nano6080144
14. Hochbaum, A.I., Chen, R., Delgado, R.D., Liang, W., Garnett, E.C., Najarian, M., Majumdar, A., Yang, P.: Enhanced thermoelectric performance of rough silicon nanowires. Nature 451(7175), 163–167 (2008). https://doi.org/10.1038/nature06381
15. Boukai, A.I., Bunimovich, Y., Tahir-Kheli, J., Yu, J.K., Heath, J.R.: Silicon nanowires as efficient thermoelectric materials. Nature 451(7175), 168–171 (2008). https://doi.org/10.1038/nature06458
16. Hicks, L.D., Dresselhaus, M.S.: Thermoelectric figure of merit of a one-dimensional conductor. Phys. Rev. B 47(24), 16631 (1993). https://doi.org/10.1103/PhysRevB.47.16631
17. Wang, Y.C., Li, B.H., Xie, G.H.: Significant reduction of thermal conductivity in silicon nanowires by shell doping. RSC Adv. 3(48), 26074–26079 (2013). https://doi.org/10.1039/c3ra45113e
18. Pan, Y., Hong, G., Raja, S.N., Zimmermann, S., Tiwari, M.K., Poulikakos, D.: Significant thermal conductivity reduction of silicon nanowire forests through discrete surface doping of germanium. Appl. Phys. Lett. 106(9), 777 (2015). https://doi.org/10.1063/1.4913879
19. Zhang, T., Wu, S., Xu, J., Zheng, R., Cheng, G.: High thermoelectric figure-of-merits from large-area porous silicon nanowire arrays. Nano Energy 13, 433–441 (2015). https://doi.org/10.1016/j.nanoen.2015.03.011
20. Liu, L., Chen, X.: Effect of surface roughness on thermal conductivity of silicon nanowires. J. Appl. Phys. 107(3), 163 (2010). https://doi.org/10.1063/1.3298457
21. Fan, D., Sigg, H., Spolenak, R., Ekinci, Y.: Strain and thermal conductivity in ultrathin suspended silicon nanowires. Phys. Rev. B 96(11), 115307 (2017). https://doi.org/10.1103/PhysRevB.96.115307
22. Murphy, K.F., Piccione, B., Zanjani, M.B., Lukes, J.R., Gianola, D.S.: Strain- and defect-mediated thermal conductivity in silicon nanowires. Nano Lett. 14(7), 3785–3792 (2014). https://doi.org/10.1021/nl500840d

23. Li, H.P., Zhang, R.Q.: Anomalous effect of hydrogenation on phonon thermal conductivity in thin silicon nanowires. EPL **105**(5), 56003 (2014). https://doi.org/10.1209/0295-5075/105/56003

24. Li, H.P., De Sarkar, A., Zhang, R.Q.: Surface-nitrogenation-induced thermal conductivity attenuation in silicon nanowires. EPL **96**(5), 56007 (2011). https://doi.org/10.1209/0295-5075/96/56007

25. Zhang, Z.W., Chen, J.: Thermal conductivity of nanowires. Chin. Phys. B **27**(3), 0351 (2018). https://doi.org/10.1088/1674-1056/27/3/035101

26. Chen, J., Zhang, G., Li, B.: A universal gauge for thermal conductivity of silicon nanowires with different cross sectional geometries. J. Chem. Phys. **135**(20), 204705 (2011). https://doi.org/10.1063/1.3663386

27. Chen, J., Zhang, G., Li, B.: Remarkable reduction of thermal conductivity in silicon nanotubes. Nano Lett. **10**(10), 3978–3983 (2010). https://doi.org/10.1021/nl101836z

28. Zhang, Y., Bi, K., Chen, W., Chen, M., Chen, Y.: The effects of doping pattern on lattice thermal conductivity of silicon nanowires. ECS Trans. **60**(1), 1159–1164 (2014). https://doi.org/10.1149/06001.1159ecst

29. Yang, N., Zhang, G., Li, B.: Ultralow thermal conductivity of isotope-doped silicon nanowires. Nano Lett. **8**(1), 276–280 (2008). https://doi.org/10.1021/nl0725998

30. And, N.M., Yang, L., And, D.L., Majumdar, A.: Predicting the thermal conductivity of Si and Ge nanowires. Nano Lett. **3**(12), 1713–1716 (2003). https://doi.org/10.1021/nl034721i

31. Ponomareva, I., Srivastava, D., Menon, M.: Thermal conductivity in thin silicon nanowires: phonon confinement effect. Nano Lett. **7**(5), 1155 (2007). https://doi.org/10.1021/nl062823d

32. Kwon, S., Wingert, M.C., Zheng, J., Xiang, J., Chen, R.: Thermal transport in Si and Ge nanostructures in the confinement regime. Nanoscale **8**(27), 13155–13167 (2016). https://doi.org/10.1039/C6NR03634A

33. Markussen, T., Jauho, A.P., Brandbyge, M. (2009). Surface-decorated silicon nanowires: a route to high-ZT thermoelectrics. Phys. Rev. Lett. **103**(5), 055502 (2009). https://doi.org/10.1103/PhysRevLett.103.055502

34. Dmitriev, A.V., Zvyagin, I.P.: Current trends in the physics of thermoelectric materials. Phys. Usp. **53**(8), 821–838 (2010). https://doi.org/10.3367/UFNe.0180.201008b.0821

35. Yang, N., Xu, X., Zhang, G., Li, B.: Thermal transport in nanostructures. AIP Adv. **2**(4), 041410 (2012). https://doi.org/10.1063/1.4773462

36. Kurosaki, K., Yusufu, A., Miyazaki, Y., Ohishi, Y., Muta, H., Yamanaka, S.: Enhanced thermoelectric properties of silicon via nanostructuring. Mater. Trans. **57**(7), 1018–1021 (2016). https://doi.org/10.2320/matertrans.MF201601

37. Zhang, G., Zhang, Y.W.: Thermal conductivity of silicon nanowires: from fundamentals to phononic engineering. Phys. Stauts Solidi RRL **7**(10), 754–766 (2013). https://doi.org/10.1002/pssr.201307188

38. Li, H.P., Zhang, R.Q.: Surface effects on the thermal conductivity of silicon nanowires. Chin. Phys. B **27**(3), 036801 (2018). https://doi.org/10.1088/1674-1056/27/3/036801

39. Zeng, Y.J., Liu, Y.Y., Zhou, W.X., Chen, K.Q.: Nanoscale thermal transport: theoretical method and application. Chin. Phys. B **27**(3), 036304 (2018). https://doi.org/10.1088/1674-1056/27/3/036304

40. Ali, A., Chen, Y., Vasiraju, V., Vaddiraju, S.: Nanowire-based thermoelectrics. Nanotechnology **28**(28), 282001 (2017). https://doi.org/10.1088/1361-6528/aa75ae

41. Schierning, G.: Silicon nanostructures for thermoelectric devices: a review of the current state of the art. Phys. Status Solid A **211**(6), 1235–1249 (2014). https://doi.org/10.1002/pssa.201300408

42. Zhang, G., Li, B.: Impacts of doping on thermal and thermoelectric properties of nanomaterials. Nanoscale **2**(7), 1058–1068 (2010). https://doi.org/10.1039/c0nr00095g

43. Kremer, R.K., Graf, K., Cardona, M., Devyatykh, G.G., Gusev, A.V., Gibin, A.M., Taldenkov, A.N., Pohl, H.J.: Thermal conductivity of isotopically enriched ^{28}Si: revisited. Phys. Status Solidi C **131**(8), 499–503 (2004). https://doi.org/10.1016/j.ssc.2004.06.022

44. Volz, S.G., Chen, G.: Molecular dynamics simulation of thermal conductivity of silicon nanowires. Appl. Phys. Lett. **75**(14), 2056–2058 (1999). https://doi.org/10.1063/1.124914

45. Schelling, P.K., Phillpot, S.R., Keblinski, P.: Comparison of atomic-level simulation methods for computing thermal conductivity. Phys. Rev. B **65**(14), 114306 (2002). https://doi.org/10.1103/PhysRevB.65.144306

46. Sellan, D.P., Landry, E.S., Turney, J.E., Mcgaughey, A.J.H., Amon, C.H.: Size effects in molecular dynamics thermal conductivity predictions. Phys. Rev. B **60**(21), 515–526 (2010). https://doi.org/10.1103/PhysRevB.81.214305

47. Li, X., Maute, K., Dunn, M.L., Yang, R.: Strain effects on the thermal conductivity of nanostructures. Phys. Rev. B **81**(24), 245318 (2010). https://doi.org/10.1103/PhysRevB.81.245318

48. Donadio, D., Galli, G.: Temperature dependence of the thermal conductivity of thin silicon nanowires. Nano Lett. **10**(3), 847–851 (2010). https://doi.org/10.1021/nl903268y

49. Chen, Y., Lukes, J.R., Li, D.: Monte carlo simulation of silicon nanowire thermal conductivity. J. Heat Transfer **127**(10), 1129–1137 (2005). https://doi.org/10.1115/1.2035114

50. Lacroix, D., Joulain, K., Lemonnier, D.: Monte carlo transient phonon transport in silicon and germanium at nanoscales. Phys. Rev. B **72**(6), 19771–19778 (2005). https://doi.org/10.1103/PhysRevB.72.064305

51. Markussen, T., Jauho, A.P., Brandbyge, M.: Heat conductance is strongly anisotropic for pristine silicon nanowires. Nano Lett. **8**(11), 3771–3775 (2008). https://doi.org/10.1021/nl8020889

52. Markussen, T., Jauho, A.P., Brandbyge, M.: Electron and phonon transport in silicon nanowires: an atomistic approach to thermoelectric properties. Phys. Rev. B **79**(3), 035415 (2009). https://doi.org/10.1103/PhysRevB.79.035415

53. Mingo, N.: Calculation of nanowire thermal conductivity using complete phonon dispersion relations. Phys. Rev. B **68**(11), 845–846 (2003). https://doi.org/10.1103/PhysRevB.68.113308

54. Mingo, N., Broido, D.A.: Lattice thermal conductivity crossovers in semiconductor nanowires. Phys. Rev. Lett. **93**(24), 246106 (2004). https://doi.org/10.1103/PhysRevLett.93.246106

55. Li, D., Wu, Y., Kim, P., Shi, L., Yang, P., Majumdar, A.: Thermal conductivity of individual silicon nanowires. Appl. Phys. Lett. **83**(14), 2934–2936 (2003). https://doi.org/10.1063/1.1616981

56. Martin, P.N., Aksamija, Z., Pop, E., Ravaioli, U.: Reduced thermal conductivity in nanoengineered rough Ge and GaAs nanowires. Nano Lett. **10**(4), 1120–1124 (2010). https://doi.org/10.1021/nl902720v

57. Malhotra, A., Maldovan, M.: Impact of phonon surface scattering on thermal energy distribution of Si and SiGe nanowires. Sci. Rep. **6**, 25818 (2016). https://doi.org/10.1038/srep25818

58. Xie, G., Guo, Y., Li, B., Yang, L., Zhang, K., Tang, M., Zhang, G.: Phonon surface scattering controlled length dependence of thermal conductivity of silicon nanowires. Phys. Chem. Chem. Phys. **15**(35), 14647–14652 (2013). https://doi.org/10.1039/c3cp50969a

59. Kukita, K., Kamakura, Y.: Monte-Carlo simulation of phonon transport in silicon including a realistic dispersion relation. J. Appl. Phys. **114**(15), 258 (2013). https://doi.org/10.1063/1.4826367

60. Lacroix, D., Joulain, K., Terris, D., Lemonnier, D.: Monte carlo simulation of phonon confinement in silicon nanostructures: application to the determination of the thermal conductivity of silicon nanowires. Appl. Phys. Lett. **89**(10), 103104 (2006). https://doi.org/10.1063/1.2345598

61. Zhang, R.Q., Lifshitz, Y., Ma, D.D.D., Zhao, Y.L., Frauenheim, T., Lee, S.T., Tong, S.Y.: Structures and energetics of hydrogen-terminated Silicon nanowire surfaces. J. Chem. Phys. **123**(14), 144703 (2005). https://doi.org/10.1063/1.2047555

62. Yang, X.B., Zhang, R.Q.: Metallization induced by nitrogen atom adsorption on Silicon nanofilms and nanowires. Appl. Phys. Lett. **94**(11), 113101 (2009). https://doi.org/10.1063/1.3098455

63. Guo, C.S., Luo, L.B., Yuan, G.D., Yang, X.B., Zhang, R.Q., Zhang, W.J., Lee, S.T.: Surface passivation and transfer doping of Silicon nanowires. Angew. Chem. Int. Ed. **48**(52), 9896–9900 (2009). https://doi.org/10.1002/anie.200906383

64. Xu, H., Yang, X.B., Zhang, C., Lu, A.J., Zhang, R.Q.: Stabilizing and activating dopants in <112> silicon nanowires by alkene adsorptions: a first-principles study. Appl. Phys. Lett. **98**(7), 335 (2011). https://doi.org/10.1063/1.3557067

65. Pan, Y., Ye, T., Qin, G., Fedoryshyn, Y., Raja, S.N., Hu, M., Degen, C.L., Poulikakos, D.: Surface chemical tuning of phonon and electron transport in free–standing Silicon nanowire arrays. Nano Lett. **16**(10), 6364–6370 (2016). https://doi.org/10.1021/acs.nanolett.6b02754

66. Lim, J., Hippalgaonkar, K., Andrews, S.C., Majumdar, A., Yang, P.: Quantifying surface roughness effects on phonon transport in silicon nanowires. Nano Lett. **12**(5), 2475–2482 (2012). https://doi.org/10.1021/nl3005868

67. Moore, A.L., Saha, S.K., Prasher, R.S., Shi, L.: Phonon backscattering and thermal conductivity suppression in sawtooth nanowires. Appl. Phys. Lett. **93**(8), 1713 (2008). https://doi.org/10.1063/1.2970044

68. Wang, Z., Ni, Z., Zhao, R., Chen, M., Bi, K., Chen, Y.: The effect of surface roughness on lattice thermal conductivity of silicon nanowires. Phys. B **406**(13), 2515–2520 (2011). https://doi.org/10.1016/j.physb.2011.03.046

69. Martin, P., Aksamija, Z., Pop, E., Ravaioli, U.: Impact of phonon-surface roughness scattering on thermal conductivity of thin Si nanowires. Phys. Rev. Lett. **102**(12), 125503 (2009). https://doi.org/10.1103/PhysRevLett.102.125503

70. Kim, H., Park, Y.H., Kim, I., Kim, J., Choi, H.J., Kim, W.: Effect of surface roughness on thermal conductivity of VLS-grown rough $Si_{1-x}Ge_x$ nanowires. Appl. Phys. A **104**(1), 23–28 (2011). https://doi.org/10.1007/s00339-011-6475-0

71. Park, I., Li, Z., Pisano, A.P., Williams, R.S.: Selective surface functionalization of Silicon nanowires via nanoscale Joule heating. Nano Lett. **7**(10), 3106–3111 (2007). https://doi.org/10.1021/nl071637k

72. Lu, A.J., Zhang, R.Q., Lee, S.T.: Tunable electronic band structures of hydrogen-terminated <112> silicon nanowires. Appl. Phys. Lett. **92**(20), 203109 (2008). https://doi.org/10.1063/1.2936088

73. Scott, S.A., Peng, W., Kiefer, A.M., Jiang, H., Knezevic, I., Savage, D.E., Lagally, M.G.: Influence of surface chemical modification on charge transport properties in ultrathin Silicon membranes. ACS Nano **3**(7), 1683–1692 (2009). https://doi.org/10.1021/nn9000947

74. Chen, J., Zhang, G., Li, B.: Tunable thermal conductivity of $Si_{1-x}Ge_x$ nanowires. Appl. Phys. Lett. **95**(7), 073117 (2009). https://doi.org/10.1063/1.3212737

75. Zhong, J., Stocks, G.M.: Localization/quasi-delocalization transitions and quasi-mobility-edges in shell-doped nanowires. Nano Lett. **6**(1), 128–132 (2006). https://doi.org/10.1021/nl051981m

76. Hu, M., Giapis, K.P., Goicochea, J.V., Zhang, X., Poulikakos, D.: Significant reduction of thermal conductivity in Si/Ge core-shell nanowires. Nano Lett. **11**(2), 618–623 (2010). https://doi.org/10.1021/nl103718a

77. Liu, X., Zhang, G., Pei, Q., Zhang, Y.: Modulating the thermal conductivity of Silicon nanowires via surface amorphization. Sci. China. Technol. Sci. **57**(4), 699–705 (2014). https://doi.org/10.1007/s11431-014-5496-2

78. Tan, X.J., Liu, G.Q., Shao, H.Z., Xu, J.T., Yu, B., Jiang, H.C., Jiang, J.: Acoustic phonon softening and reduced thermal conductivity in $Mg_2Si_{1-x}Sn_x$ solid solutions. Appl. Phys. Lett. **110**(14), 143903 (2017). https://doi.org/10.1063/1.4979871

79. Wingert, M.C., Kwon, S., Hu, M., Poulikakos, D., Xiang, J., Chen, R.: Sub-amorphous thermal conductivity in ultrathin crystalline silicon nanotubes. Nano Lett. **15**(4), 2605–2611 (2015). https://doi.org/10.1021/acs.nanolett.5b00167

80. Yang, L., Yang, Y., Zhang, Q., Zhang, Y., Jiang, Y., Guan, Z., Xu, T.T.: Thermal conductivity of individual silicon nanoribbons. Nanoscale **8**(41), 17895–17901 (2016). https://doi.org/10.1039/C6NR06302K

81. Neogi, S., Reparaz, J.S., Pereira, L.F.C., Graczykowski, B., Wagner, M.R., Sledzinska, M., Donadio, D.: Tuning thermal transport in ultrathin silicon membranes by surface nanoscale engineering. ACS Nano **9**(4), 3820–3828 (2015). https://doi.org/10.1021/nn506792d
82. Massoud, A.M., Bluet, J.M., Lacatena, V., Haras, M., Robillard, J.F., Chapuis, P.O.: Native-oxide limited cross-plane thermal transport in suspended silicon membranes revealed by scanning thermal microscopy. Appl. Phys. Lett. **111**(6), 063106 (2017). https://doi.org/10.1063/1.4997914

Chapter 5
Phonon Thermal Transport in Silicene and Its Defect Effects

The shrinkage of dimensions of nanomaterials is expected to reveal new fascinating properties that may ultimately lead to exciting applications. In the nano realm, in addition to the zero-dimensional (0D) quantum dots and the one-dimensional (1D) nanowires, silicon can also appear in two-dimensional (2D) nanosheet forms [1]. 2D nanosheets bridge the gap between 1D nanomaterials and 3D macroscale bulk materials, and will advance understanding of the quantum confinement effect across the boundaries of all dimensions in materials science. The past decades witnessed a surge of research interests in the structures and properties of 2D materials in the nano realm, which has been partially boosted by the development of graphene [2], an achievement earned the Nobel Prize in physics for Geim and Novoselov in 2010. Graphene is endowed with exceptional mechanical, electronic and optical properties [3]. It also has ultrahigh thermal conductivity (in the range of 3000–5000 W/mK [4]), which is extremely useful for heat removal in microelectronics. Therefore, numerous theoretical and experimental research on graphene materials are driven by both fundamental scientific interest and the perspective of applications. Interested readers can refer to several review articles [5–9] and books [10–12] for further details.

Following the intensive research devoted to graphene, many other 2D materials have been investigated to determine if they have potential analogous to graphene's characteristics. Silicon's counterpart of graphene, named silicene [13], is focused on because of its good compatibility with existing silicon-based electronics. Silicene, a new 2D material, is a recently-observed monolayer of silicon atoms arranged in a honeycomb lattice structure. Benefiting from the advancement in nanotechnology, silicene nanoribbons (SNRs) have been synthesized on metal substrates, such as Ag(111) [14, 15], Ag(110) [16], Au(110) [17], Ir(111) [18] and ZrB_2(0001) [19] and are predicted to be stable on non-metallic substrates [20]. Despite the difficulties in synthesizing free-standing silicene thus far, such experimental progress paved the way for the synthesis of free-standing silicene, which is attracting intense interest from the scientific community.

© The Author(s), under exclusive licence to Springer Nature Singapore Pte Ltd. 2018   67
H.-P. Li and R.-Q. Zhang, *Phonon Thermal Transport in Silicon-Based Nanomaterials*,
SpringerBriefs in Physics, https://doi.org/10.1007/978-981-13-2637-0_5

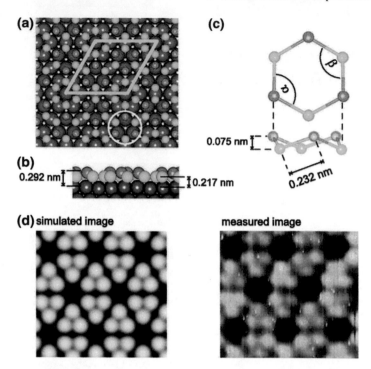

Fig. 5.1 Density functional theory results for silicene on Ag(111). **a** Top view and **b** side view of the fully relaxed atomic geometries of the model for silicene on the Ag(111) surface. **c** Enlarged image of the hexagonal silicene ring indicated by the white circle in (**a**). **d** Simulated STM image (left) for the structure shown in (**a**). The simulated image exhibits the same structural features as those observed in the experimental STM image (right), i.e., a hexagonal arrangement of the triangular structure around a dark center. Reprinted with permission from Ref. [14], copyright (2012) by the American Physical Society

Both theoretical [13, 21, 22] and experimental [14–16, 23] studies show that the honeycomb lattice of silicene is slightly buckled (Fig. 5.1) in its most stable form, which differs from the known planar structure of graphene. Unlike carbon atoms in graphene, silicon atoms tend to adopt sp^3 hybridization over sp^2 in silicene [24], which makes it highly chemically active on the surface and allows its electronic states to be easily tuned by chemical functionalization [25]. Its buckled structure also gives silicene a tunable band gap by applying an external electric field [26].

Density functional theory calculations and experimental observations indicate that buckled silicene has electronic properties similar to those of graphene, such as Dirac cone [27], massless Dirac fermions [28], and quantum spin Hall effect [29]. Silicene has great potential to be used in the next generation of electronics with the advantage of being compatible with existing silicon-based electronics. Compared to the extensive studies on the structural and electronic properties of silicene, the thermal property of silicene remains largely unexplored. The thermal conductivity of silicene is remarkably significant for thermal management and thermoelectric conversion in silicene-based electronic devices. Therefore, investigating the thermal conductivity of silicene is of both academic and practical importance. In this chapter, we briefly review the current advances in phonon thermal transport of silicene and its nanoribbons, and particularly highlight our recent theoretical work [30, 31], including the vacancy defect effect and isotope defect effect on phonon thermal conductivity of silicene system, respectively.

5.1 Phonon Thermal Transport in Pristine Silicene and Silicene Nanoribbons

The buckling in silicene leads to new properties [26, 32] which make it a promising alternative to graphene in the rapidly developing areas of thermoelectrics [33] and nanoelectronics [34]. Silicene possesses better electronic properties than graphene; but its thermal transport properties have not been fully studied [35, 36]. As mentioned, graphene exhibits high phonon thermal conductivity which may be useful in applications such as electronic cooling and heat dissipation. However, high thermal conductivity is undesirable in thermoelectric materials where an extremely low thermal conductivity is required. Unlike graphene, silicene has a smaller lattice thermal conductivity and can be effectively used in thermoelectricity in future. Given the difficulties in synthesizing free-standing silicene, the thermal conductivity of silicene has not been experimentally measured yet, and only few results exist instead on the phonon thermal conductivity of silicene using molecular dynamics (MD) simulations [30, 31, 37–41] and first-principles calculations [36, 42–45]. To our knowledge, we were first to report the phonon thermal conductivity of perfect silicene in [30]. Using equilibrium MD simulations, we predicted the overall phonon thermal conductivity around 20 W/mK at room temperature for a silicene sheet [30], which is much lower than that of bulk silicon (~150 W/mK). Subsequently, non-equilibrium MD methods, which rigorously rely on the empirical interatomic potentials, were applied to calculate room-temperature thermal conductivity of silicene and predict the values of thermal conductivity in the range of 5–69 W/mK [37–41]. Notably, first-principles-based lattice dynamics predicted that the room-temperature thermal conductivity of silicene ranges from 20 to 30 W/mK [42–45], which should be more reliable. In the first-principles approach, the harmonic second-order and anharmonic third-order interatomic force constants [46–49] are determined from density functional perturbation theory as input parameters; then, the phonon Boltzmann transport equation is solved [44, 50].

More recently, Li and coworkers [31] have investigated the length dependence and edge-chirality dependence of phonon thermal conductivity in SNRs using non-equilibrium MD simulations. They found that the thermal conductivity of armchair-edged SNR is 10% smaller than that of zigzag-edged SNR, similar to previously reported results [51] and in good agreement with our previous equilibrium MD result of $\kappa_{\text{zigzag}}/\kappa_{\text{armchair}} \sim 1.09 \pm 0.11$ [30]. This phenomenon implies that silicene demonstrates slight chirality dependence in heat conduction. Such anisotropic behavior can be qualitatively explained by the different mean free path (MFP) of phonons along two chiral directions. For the same system length L, the phonon MFPs along the armchair and zigzag directions are estimated to be $1.35L$ and $1.15L$, respectively [51, 52]. Thus, longer MFP phonons restricted on a nanoscale can be remarkably scattered by the boundary, resulting in lower thermal conductivity.

Consequently, low thermal conductivity has been estimated for silicene nanosheet, and this feature could be exploited for thermoelectric applications. Calculations of the thermoelectric properties of armchair and zigzag SNRs suggest that this material may be attractive for thermoelectric devices [33, 53, 54]. The key to improving the thermoelectric efficiency of silicene relies on opening a bandgap to enhance the thermopower and suppress lattice thermal conductivity. Given their buckled atomic structure, SNRs have a nonzero energy gap, which can be tuned further by applying external transverse electric fields and doping [26, 53]. Several strategies have also been utilized to reduce the thermal conductivity in SNRs, including adding substrates [55, 56], heterostructuring [57], surface functionalization [41], and defects [31, 51, 53, 58, 59]. These studies provide significant guidance for experimental realization.

5.2 Phonon Thermal Transport in Vacancy-Defected Silicene

As well known, defects or impurities are ubiquitous in natural materials. As working with defect-free or impurity-free materials is almost impossible, understanding how defects and impurities alter the electronic and thermal properties of systems is essential [60]. Structural defects, such as the Stone-Wales and vacancy defects, are inevitable in the synthesis of 2D materials with hexagonal lattice structure [61]. Furthermore, physical methods, such as stress, irradiation, and sublimation, can also generate a considerable concentration of such defects [31, and references therein]. Vacancy defects are lattice sites that would otherwise be occupied in a perfect crystal, but instead remain vacant. Such defects significantly affect the mechanical properties [61, 62], electronic properties [63, 64] and thermal stability [65] of silicene sheets, and cause lattice vibration localized around the defects. Recent studies have shown that vacancies can lower phonon thermal conduction in graphene [66, 67] and in silicon nanowires [68] as well as in silicene in our prior study [30].

We investigated the effects of the concentration of monoatomic vacancies and the size of vacancy clusters on the in-plane thermal conductivity of silicene sheets [30], as shown in Fig. 5.2. Even for the lowest monovacancy concentration ρ of 0.22% (i.e., one of the 448 atoms was removed from the model sheet), thermal conductivity reduction can be remarkable (by 78%) compared with a perfect case. Fitting the simulation results in Fig. 5.2a gives $\kappa / \kappa_0 = 1 / (1.166 + 1.529\rho)$, indicating a significant reduction in thermal conductivity at ρ below 2%, followed by a slow decrease in thermal conductivity at higher monovacancy concentrations. In addition, our simulations

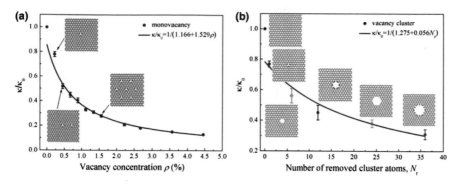

Fig. 5.2 Normalized in-plane thermal conductivity (κ/κ_0) of silicene sheets as a function of **a** the monoatomic vacancy concentration and **b** the number of cluster atoms removed. Grey balls denote silicon atoms, whereas other balls indicate the edge atoms of the vacancies. Reprinted with permission from Ref. [30], copyright (2012) by EPLA

[30] on the thermal conductivity of silicene sheets with different-sized vacancy clusters revealed that vacancy clusters degrade the thermal conductivity more remarkably than monoatomic vacancies do, as shown in Fig. 5.2b. In particular, removing a 12-atom cluster decreases the thermal conductivity of a pristine sheet by half, whereas a monoatomic vacancy induces a reduction of approximately 78%. Increasing the size of a vacancy cluster thus results in a larger reduction in the thermal conductivity owing to the stronger effect of phonon-defect scattering.

We also elucidated the thermal conductivity reduction in silicene through analysis of the in-plane phonon density of state (PDOS) spectra associated with vacancy defects in silicene sheets [30]. Using equilibrium MD simulations, the PDOS (or the phonon vibrational power spectrum) is computed from the Fourier transform of the velocity autocorrelation function (VAF) [69], PDOS $= \frac{1}{\sqrt{2\pi}} \int_0^\infty e^{i\omega t} C_{vv}(t) dt$ where the VAF is defined as $C_{vv}(t) = \left\langle \sum_j v_j(t_0 + t) \cdot v_j(t_0) \right\rangle$. Here $v_j(t)$ is the velocity vector of atom j at time t, the summations are over the atoms in the relevant part of interest, and the brackets indicate ensemble averaging over different time origins t_0. When the x, y, or z component of PDOS is considered, the projected PDOS is calculated by taking the corresponding component of the velocity vector. Phonon modes with frequencies below 20 THz dominate the thermal transport in silicon [70], and the simulated PDOS for bulk silicon shows excellent agreement with the experimental and theoretical data [30, and references therein]. Compared with bulk silicon, the in-plane PDOS for pristine silicene sheets exhibits a significant depression in the considered frequency region because of the effect of surface-phonon scattering. Additionally, our simulations [30] reveal a shortening of the Si–Si bond for free surfaces owing to bond reconstruction; thus, a stiffer surface bond could account for a significant shift in the entire PDOS to higher frequencies. The depression of the PDOS reduces the density of thermal energy carriers [71], and the blue-shift of a low-frequency PDOS may decrease the contribution of long-wavelength phonon modes

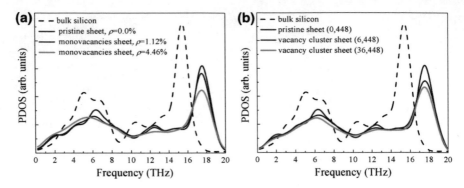

Fig. 5.3 In-plane PDOS analysis of the effect of **a** the monoatomic vacancy concentration, and **b** the size of a vacancy cluster for a silicene sheet. ρ is the concentration of monoatomic vacancies. (6, 448) denotes that a 6-atom cluster was removed from a 448-atom cell. Curves obtained from MD simulations are drawn by a Gaussian smoothing with a width of 1.2 THz. Reprinted with permission from Ref. [30], copyright (2012) by EPLA

dominating the heat conduction at low temperature [70], both of which contribute directly to thermal conductivity reduction.

We previously investigated the effect of vacancy defects on the in-plane PDOS of silicene sheets [30]. As shown in Fig. 5.3, by increasing the concentration of monovacancies or the size of the vacancy cluster, the high frequency parts of the PDOS near 18 THz are remarkably reduced, leading to lower thermal conductivity, as the optical modes could also carry heat [72]. Furthermore, a remarkable broadening of the acoustic phonon modes near 6 THz and 12.5 THz occurs as vacancy concentration and hole size increase. Particularly for larger contractions or larger sized vacancy defects, the valleys and peaks in the PDOS curve between 10 THz and 16 THz almost disappear, and this part of the PDOS curve flattens. Such broadening of the PDOS upon introducing vacancies has also been observed experimentally in $La_{3-x}Te_4$ [73] and theoretically in graphene [74], indicating a reduction in the lifetime (or MFP) of the related phonon modes. Vacancies in a silicon nanosheet strongly scatter the phonons, and this source of scattering is associated with the concentration and size of the vacancies. Therefore, by introducing vacancy defects, certain acoustic vibration modes become localized and thus adversely contribute to thermal conductivity.

The boundary shape of the vacancy clusters plays an important role in the thermal conductivity of silicene sheets [30]. As shown in Fig. 5.4, phonon thermal transport in the defective sheets exhibits unique features associated with the boundary shape and orientation of the vacancy clusters, unlike those from the pristine sheet. Our investigation [30] shows a significant anisotropy of the phonon thermal conductivity in the vacancy-cluster-defected silicene sheet. In particular, the value of κ_x is roughly 1.3 times larger than the κ_y value in the R1 orientation, but the former is approximately half of the latter in the R2 orientation. Such remarkable anisotropy of thermal conductivity induced by vacancy defects could have potential applications, including

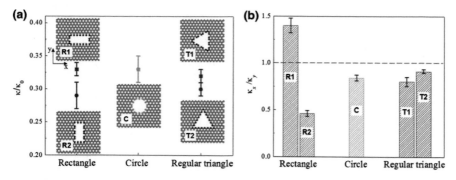

Fig. 5.4 **a** Normalized in-plane thermal conductivity (κ/κ_0) and **b** thermal conductivity anisotropy ratio (κ_x/κ_y) of silicene sheets with different boundary shapes of vacancy clusters at 300K. The red, green, and blue balls denote the edge silicon atoms of rectangular vacancy clusters (R1, R2), a circular vacancy cluster (C), and regular-triangular vacancy clusters (T1, T2), respectively. Here the vacancy clusters are formed by the removal of Si_{36} from the 14 × 8 cell with 448 atoms. $\kappa_x/\kappa_y=1$ corresponds to the pure isotropy of the phonon thermal conductivity. Reprinted with permission from Ref. [30], copyright (2012) by EPLA

improving the efficiency of thermoelectric converters and designing phonon logic devices [75].

5.3 Phonon Thermal Transport in Isotope-Doped Silicene Nanoribbons

Isotope impurities provide a powerful technique for modulating the phonon-related properties of nanomaterials. Technologically speaking, isotope doping easily introduces phonon-defect scattering without damaging electronic quality, thus attracting intensive research interests in 1D nanostructures, such as nanowires [74] and nanotubes [76–79], and 2D nanostructures including graphene [80, 81]. However, the influence of isotope doping, including doping concentration, doping pattern, and doping type on the thermal conductivity of SNRs, remains unclear. Given the extreme sensitivity of electron transport in nanostructures to isotope doping [82, 83], a parallel, simultaneous investigation of phonon transport is clearly needed [84].

Using non-equilibrium MD simulation, Li et al. [31] have recently introduced the isotope to modulate the phonon thermal conductivity of SNRs with two patterns: random atomic distribution of isotopes and superlattice-structured isotope substitution. Figure 5.5 displays the calculated thermal conductivity of randomly doped armchair-edged SNR as a function of the ^{30}Si doping concentration (ρ). The mean field model [85] results are in good agreement with the MD calculations. A U-shaped change of the thermal conductivity in the studied SNR is also observed. Phonon thermal conductivity decreases initially to a minimum and then increases as the doping concentration changes from 0 to 100%, and the minimum (about 15.5 W/mK) occurs at

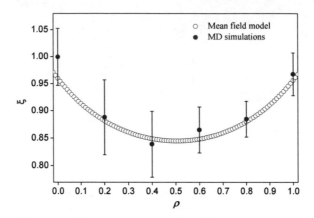

Fig. 5.5 Plot of the phonon thermal conductivity reduction ratio ξ against the doping concentration ρ for ^{30}Si randomly doped armchair-edged SNR, as calculated with non-equilibrium MD simulations (blue solid circle) and mean field model (red circle). Here, thermal conductivities are normalized by the phonon thermal conductivity of pure ^{28}Si SNR. The armchair-edged SNR is 30 nm in length and 3 nm in width. Reprinted with permission from Ref. [31], copyright (2018) by the Chinese Physical Society and IOP Publishing

the doping concentration of approximately 50%. Specifically, the calculated thermal conductivity of pure ^{30}Si SNR ($\rho = 100\%$) is found to be approximately 3.8% smaller than that of pure ^{28}Si SNR ($\rho = 0\%$, $\kappa = 18.3$ W/mK) because heavier isotope atoms decrease the lattice vibration frequencies, thus resulting in lower phonon thermal conductivity. A similar phenomenon is reported in isotope-doped silicon nanowires [82] and graphene [86]. The isotope-doping-induced reduction of phonon thermal conductivity arises from a point defect induced by a mass difference in a crystal lattice that causes phonon scattering; moreover, this finding is controlled by the masses and concentrations of the individual isotopes that contribute to the disorder [37].

Li et al. [31] further studied the mechanism of thermal conductivity reduction induced by isotopic doping. As shown in Fig. 5.6, with the increase of doping concentration ρ from 0 to 100%, the frequencies of the phonon modes show redshift. When ρ increases from 0 to 40% or decreases from 100 to 60%, the mass disorder increases, thereby lowering the intensity of the PDOS peaks. Doping isotope also softens the G-band of the phonons compared with the pure system. Isotope substitution induces mass disorder in the lattice, resulting in increased phonon scattering at defective sites because of the difference in the characteristic frequencies of phonons. Those impurities cause localization of phonon modes, thus reducing thermal conductivity.

Moreover, Li et al. [31] explored the effect of ordered manner, i.e., in a superlattice arrangement (Fig. 5.7a), on the thermal conductivity of SNR. Figure 5.7b shows the phonon thermal conductivity of superlattice-structured SNR as a function of period length L_P for doping concentration $\rho = 40\%$. When L_P increases gradually from 3 to 10 nm, the κ value of the studied superlattice-structured SNR decreases

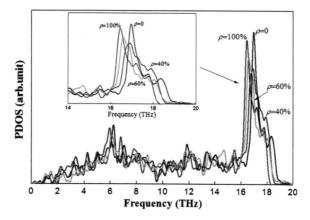

Fig. 5.6 PDOS of randomly doped armchair-edged SNR with different doping concentrations ρ of
^{30}Si. Reprinted with permission from Ref. [31], copyright (2018) by the Chinese Physical Society
and IOP Publishing

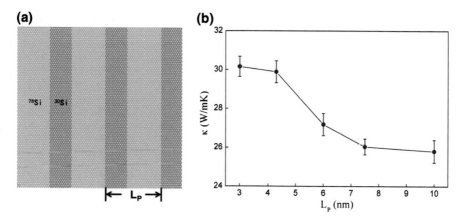

Fig. 5.7 **a** Schematic of the superlattice-structured SNR of 30 nm in length and 30 nm in width. **b**
Thermal conductivity of superlattice-structured SNR as a function of period length L_P for doping
concentration ρ = 40%. Heat flows perpendicularly to the doping strip region. Reprinted with
permission from Ref. [31], copyright (2018) by the Chinese Physical Society and IOP Publishing

from 31 (L_P = 3 nm) to 27 W/mK (L_P = 7.5 nm) and to 25 W/mK (L_P = 10
nm). This outcome can be explained qualitatively through the phonon spectrum
variations induced by superlattice arrangements [31]. Interestingly, for the same
doping concentration ρ = 40%, ordered doping (i.e., isotope superlattice) leads to a
much larger reduction in phonon thermal conductivity than random doping does, as
shown in Fig. 5.8.

Fig. 5.8 Phonon thermal conductivity of random isotope substitution and superlattice isotope substitution in SNR for the same concentration ρ = 40% compared with the case of pure SNRs (ρ = 0% and ρ = 100%). Reprinted with permission from Ref. [31], copyright (2018) by the Chinese Physical Society and IOP Publishing

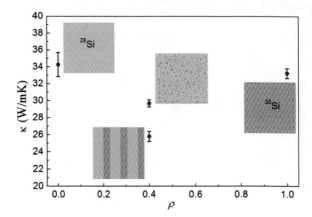

References

1. Zhang, R.Q.: Growth mechanisms and novel properties of silicon nanostructures from quantum-mechanical calculations. Springer, Berlin, Heidelberg (2014)
2. Novoselov, K.S., Geim, A.K., Morozov, S.V., Jiang, D., Zhang, Y., Dubonos, S.V., Firsov, A.A.: Electric field effect in atomically thin carbon films. Science **306**(5696), 666–669 (2004). https://doi.org/10.1126/science.1102896
3. Novoselov, K.S., Fal, V.I., Colombo, L., Gellert, P.R., Schwab, M.G., Kim, K.: A roadmap for graphene. Nature **490**(7419), 192 (2012). https://doi.org/10.1038/nature11458
4. Ghosh, D., Calizo, I., Teweldebrhan, D., Pokatilov, E.P., Nika, D.L., Balandin, A.A., Lau, C.N.: Extremely high thermal conductivity of graphene: prospects for thermal management applications in nanoelectronic circuits. Appl. Phys. Lett. **92**(15), 151911 (2008). https://doi.org/10.1063/1.2907977
5. Neto, A.H.C., Guinea, F., Peres, N.M.R., Novoselov, K.S., Geim, A.K.: The electronic properties of graphene. Rev. Mod. Phys. **81**(1), 109–162 (2009). https://doi.org/10.1103/RevModPhys.81.109
6. Allen, M.J., Tung, V.C., Kaner, R.B.: Honeycomb carbon: a review of graphene. Chem. Rev. **110**(1), 132–145 (2009). https://doi.org/10.1021/cr900070d
7. Ghuge, A.D., Shirode, A.R., Kadam, V.J.: Graphene: a comprehensive review. Curr. Drug Targets **18**(6), 724–733 (2017). https://doi.org/10.2174/1389450117666160709023425
8. Xu, X., Chen, J., Li, B.: Phonon thermal conduction in novel 2D materials. J. Phys. Cond. Matt. **28**(48), 483001 (2016). https://doi.org/10.1088/0953-8984/28/48/483001
9. Balandin, A.A.: Thermal properties of graphene and nanostructured carbon materials. Nature Mater. **10**(8), 569–581 (2011). https://doi.org/10.1038/nmat3064
10. Zhang, G. (ed.): Thermal Transport in Carbon-Based Nanomaterials. Elsevier, New York (2017)
11. Zhang, G. (ed.): Nanoscale Energy Transport and Harvesting: A Computational Study. Pan Stanford, Singapore (2015)
12. Shafraniuk, S.: Graphene: Fundamentals, Devices and Applications. Pan Stanford, Singapore (2015)
13. Cahangirov, S., Topsakal, M., Aktürk, E., Şahin, H., Ciraci, S.: Two- and one-dimensional honeycomb structures of silicon and germanium. Phys. Rev. Lett. **102**(23), 236804 (2009). https://doi.org/10.1103/PhysRevLett.102.236804
14. Vogt, P., De Padova, P., Quaresima, C., Avila, J., Frantzeskakis, E., Asensio, M.C., Le Lay, G.: Silicene: compelling experimental evidence for graphene like two-dimensional silicon. Phys. Rev. Lett. **108**(15), 155501 (2012). https://doi.org/10.1103/PhysRevLett.108.155501

15. Feng, B., Ding, Z., Meng, S., Yao, Y., He, X., Cheng, P., Wu, K.: Evidence of silicene in honeycomb structures of silicon on Ag (111). Nano Lett. **12**(7), 3507–3511 (2012). https://doi.org/10.1021/nl301047g
16. Aufray, B., Kara, A., Vizzini, S., Oughaddou, H., Leandri, C., Ealet, B., Le Lay, G.: Graphene-like silicon nanoribbons on Ag (110): a possible formation of silicene. Appl. Phys. Lett. **96**(18), 183102 (2010). https://doi.org/10.1063/1.3419932
17. Tchalala, M.R., Enriquez, H., Mayne, A.J., Kara, A., Roth, S., Silly, M.G., Dujardin, G.: Formation of one-dimensional self-assembled silicon nanoribbons on Au (110) (2 × 1). Appl. Phys. Lett. **102**(8), 083107 (2013). https://doi.org/10.1063/1.4793536
18. Meng, L., Wang, Y., Zhang, L., Du, S., Wu, R., Li, L., Gao, H.J.: Buckled silicene formation on Ir (111). Nano Lett. **13**(2), 685–690 (2013). https://doi.org/10.1021/nl304347w
19. Fleurence, A., Friedlein, R., Ozaki, T., Kawai, H., Wang, Y., Yamada-Takamura, Y.: Experimental evidence for epitaxial silicene on diboride thin films. Phys. Rev. Lett. **108**(24), 245501 (2012). https://doi.org/10.1103/PhysRevLett.108.245501
20. Kokott, S., Pflugradt, P., Matthes, L., Bechstedt, F.: Nonmetallic substrates for growth of silicene: an ab initio prediction. J. Phys.: conden. Matter **26**(18), 185002 (2014). https://doi.org/10.1088/0953-8984/26/18/185002
21. Guzmán-Verri, G.G., Voon, L.L.Y.: Electronic structure of silicon-based nanostructures. Phys. Rev. B **76**(7), 075131 (2007). https://doi.org/10.1103/PhysRevB.76.075131
22. Şahin, H., Cahangirov, S., Topsakal, M., Bekaroglu, E., Akturk, E., Senger, R.T., Ciraci, S.: Monolayer honeycomb structures of group-IV elements and III-V binary compounds: first-principles calculations. Phys. Rev. B **80**(15), 155453 (2009). https://doi.org/10.1103/PhysRevB.80.155453
23. Lin, C.L., Arafune, R., Kawahara, K., Tsukahara, N., Minamitani, E., Kim, Y., Kawai, M. (2012). Structure of silicene grown on Ag (111). Appl. Phys. Exp. **5**(4), 045802 (2010). https://doi.org/10.1143/APEX.5.045802
24. Fagan, S.B., Baierle, R.J., Mota, R., da Silva, A.J., Fazzio, A.: Ab initio calculations for a hypothetical material: silicon nanotubes. Phys. Rev. B **61**(15), 9994 (2000). https://doi.org/10.1103/PhysRevB.61.9994
25. Du, Y., Zhuang, J., Liu, H., Xu, X., Eilers, S., Wu, K., Peleckis, G.: Tuning the band gap in silicene by oxidation. ACS Nano **8**(10), 10019–10025 (2014). https://doi.org/10.1021/nn504451t
26. Drummond, N.D., Zolyomi, V., Fal'Ko, V.I.: Electrically tunable band gap in silicene. Phys. Rev. B **85**(7), 075423 (2012). https://doi.org/10.1103/PhysRevB.85.075423
27. Feng, Y., Liu, D., Feng, B., Liu, X., Zhao, L., Xie, Z., He, S.: Direct evidence of interaction-induced Dirac cones in a monolayer silicene/Ag(111) system. Proc. Natl. Acad. Sci. U.S.A. **113**(51), 14656–14661 (2016). https://doi.org/10.1073/pnas.1613434114
28. Chen, L., Liu, C.C., Feng, B., He, X., Cheng, P., Ding, Z., Wu, K.: Evidence for dirac fermions in a honeycomb lattice based on silicon. Phys. Rev. Lett. **109**(5), 056804 (2012). https://doi.org/10.1103/PhysRevLett.109.056804
29. Liu, C.C., Feng, W., Yao, Y.: Quantum spin Hall effect in silicene and two-dimensional germanium. Phys. Rev. Lett. **107**(7), 076802 (2011). https://doi.org/10.1103/PhysRevLett.107.076802
30. Li, H.P., Zhang, R.Q.: Vacancy-defect-induced diminution of thermal conductivity in silicene. EPL **99**(3), 36001 (2012). https://doi.org/10.1209/0295-5075/99/36001
31. Xu, R.F., Han, K., Li, H.P.: Effect of isotope doping on phonon thermal conductivity of silicene nanoribbons: a molecular dynamics study. Chin. Phys. B **27**(2), 026801 (2018). https://doi.org/10.1088/1674-1056/27/2/026801
32. Chowdhury, S., Jana, D.: A theoretical review on electronic, magnetic and optical properties of silicene. Rep. Prog. Phys. **79**, 126501 (2016). https://doi.org/10.1088/0034-4885/79/12/126501
33. Zberecki, K., Wierzbicki, M., Barnaś, J., Swirkowicz, R.: Thermoelectric effects in silicene nanoribbons. Phys. Rev. B **88**(11), 115404 (2013). https://doi.org/10.1103/PhysRevB.88.115404

34. Tao, L., Cinquanta, E., Chiappe, D., Grazianetti, C., Fanciulli, M., Dubey, M., Akinwande, D.: Silicene field-effect transistors operating at room temperature. Nature Nano. **10**(3), 227–231 (2015). https://doi.org/10.1038/nnano.2014.325
35. Feyzi, A., Chegel, R.: Heat capacity, electrical and thermal conductivity of silicene. Eur. Phys. J. B **89**(9), 193 (2016). https://doi.org/10.1140/epjb/e2016-70333-x
36. Gori, P., Pulci, O., de Lieto Vollaro, R., Guattari, C.: Thermophysical properties of the novel 2D materials graphene and silicene: insights from ab-initio calculations. Energ. Procedia **45**, 512–517 (2014). https://doi.org/10.1016/j.egypro.2014.01.055
37. Pei, Q.X., Zhang, Y.W., Sha, Z.D., Shenoy, V.B.: Tuning the thermal conductivity of silicene with tensile strain and isotopic doping: a molecular dynamics study. J. Appl. Phys. **114**(3), 033526 (2013). https://doi.org/10.1063/1.4815960
38. Hu, M., Zhang, X., Poulikakos, D.: Anomalous thermal response of silicene to uniaxial stretching. Phys. Rev. B **87**(19), 195417 (2013). https://doi.org/10.1103/PhysRevB.87.195417
39. Ng, T.Y., Yeo, J., Liu, Z.: Molecular dynamics simulation of the thermal conductivity of shorts strips of graphene and silicene: a comparative study. Int. J. Mech. Mater. Des. **9**(2), 105–114 (2013). https://doi.org/10.1007/s10999-013-9215-0
40. Zhang, X., Xie, H., Hu, M., Bao, H., Yue, S., Qin, G., Su, G.: Thermal conductivity of silicene calculated using an optimized Stillinger-Weber potential. Phys. Rev. B **89**(5), 054310 (2014). https://doi.org/10.1103/PhysRevB.89.054310
41. Liu, Z.Y., Wu, X.F., Luo, T.F.: The impact of hydrogenation on the thermal transport of silicene. 2D Mater. **4**(2), 025002 (2017). https://doi.org/10.1088/2053-1583/aa533e
42. Gu, X., Yang, R.: First-principles prediction of phononic thermal conductivity of silicene: a comparison with graphene. J. Appl. Phys. **117**(2), 025102 (2015). https://doi.org/10.1063/1.4905540
43. Kuang, Y.D., Lindsay, L., Shi, S.Q., Zheng, G.P.: Tensile strains give rise to strong size effects for thermal conductivities of silicene, germanene and stanene. Nanoscale **8**(6), 3760–3767 (2016). https://doi.org/10.1039/C5NR08231E
44. Xie, H., Ouyang, T.: Éric Germaneau, Qin, G., Hu, M., Bao, H.: Unexpectedly large tunability of lattice thermal conductivity of monolayer silicene via mechanical strain. Phys. Rev. B **93**(7), 075404 (2016). https://doi.org/10.1103/PhysRevB.93.075404
45. Peng, B., Zhang, H., Shao, H., Xu, Y., Zhang, R., Lu, H., Zhu, H.: First-principles prediction of ultralow lattice thermal conductivity of dumbbell silicene: a comparison with low-buckled silicene. ACS Appl. Mater. Interfaces. **8**(32), 20977–20985 (2016). https://doi.org/10.1021/acsami.6b04211
46. Ward, A., Broido, D.A., Stewart, D.A., Deinzer, G.: Ab initio theory of the lattice thermal conductivity in diamond. Phys. Rev. B **80**(12), 125203 (2009). https://doi.org/10.1103/PhysRevB.80.125203
47. Esfarjani, K., Chen, G., Stokes, H.T.: Heat transport in silicon from first-principles calculations. Phys. Rev. B **84**(8), 085204 (2011). https://doi.org/10.1103/PhysRevB.84.085204
48. Li, W., Lindsay, L., Broido, D.A., Stewart, D.A., Mingo, N.: Thermal conductivity of bulk and nanowire $Mg_2Si_xSn_{1-x}$ alloys from first principles. Phys. Rev. B **86**(17), 174307 (2012). https://doi.org/10.1103/PhysRevB.86.174307
49. Garg, J., Chen, G.: Minimum thermal conductivity in superlattices: a first-principles formalism. Phys. Rev. B **87**(14), 140302 (2013). https://doi.org/10.1103/PhysRevB.87.140302
50. Xie, H., Hu, M., Bao, H.: Thermal conductivity of silicene from first-principles. Appl. Phys. Lett. **104**(13), 131906 (2014). https://doi.org/10.1063/1.4870586
51. Liu, B., Reddy, C.D., Jiang, J., Zhu, H., Baimova, J.A., Dmitriev, S.V., Zhou, K.: Thermal conductivity of silicene nanosheets and the effect of isotopic doping. J. Phys. D Appl. Phys. **47**(16), 165301 (2014). https://doi.org/10.1088/0022-3727/47/16/165301
52. Regner, K.T., Sellan, D.P., Su, Z., Amon, C.H., McGaughey, A.J., Malen, J.A.: Broadband phonon mean free path contributions to thermal conductivity measured using frequency domain thermoreflectance. Nature Comm. **4**, 1640 (2013). https://doi.org/10.1038/ncomms2630
53. Sadeghi, H., Sangtarash, S., Lambert, C.J.: Enhanced thermoelectric efficiency of porous silicene nanoribbons. Sci. Rep. **5**, 9514 (2015). https://doi.org/10.1038/srep09514

54. Pan, L., Liu, H.J., Tan, X.J., Lv, H.Y., Shi, J., Tang, X.F., Zheng, G.: Thermoelectric properties of armchair and zigzag silicene nanoribbons. Phys. Chem. Chem. Phys. **14**(39), 13588–13593 (2012). https://doi.org/10.1039/C2CP42645E

55. Wang, Z., Feng, T., Ruan, X.: Thermal conductivity and spectral phonon properties of free-standing and supported silicene. J. Appl. Phys. **117**(8), 084317 (2015). https://doi.org/10.1063/1.4913600

56. Zhang, X., Bao, H., Hu, M.: Bilateral substrate effect on the thermal conductivity of two-dimensional silicon. Nanoscale **7**(14), 6014–6022 (2015). https://doi.org/10.1039/C4NR06523A

57. Wang, X., Hong, Y., Chan, P.K., Zhang, J.: Phonon thermal transport in silicene-germanene superlattice: a molecular dynamics study. Nanotechnology **28**(25), 255403 (2017). https://doi.org/10.1088/1361-6528/aa71fa

58. Zhao, W., Guo, Z.X., Zhang, Y., Ding, J.W., Zheng, X.J.: Enhanced thermoelectric performance of defected silicene nanoribbons. Solid State Commun. **227**, 1–8 (2016). https://doi.org/10.1016/j.ssc.2015.11.012

59. Wirth, L.J., Osborn, T.H., Farajian, A.A.: Resilience of thermal conductance in defected graphene, silicene, and boron nitride nanoribbons. Appl. Phys. Lett. **109**(17), 173102 (2016). https://doi.org/10.1063/1.4965294

60. Araujo, P.T., Terrones, M., Dresselhaus, M.S.: Defects and impurities in graphene-like materials. Mater. Today **15**(3), 98–109 (2012). https://doi.org/10.1016/S1369-7021(12)70045-7

61. Yuan, X., Lin, G., Wang, Y.: Mechanical properties of armchair silicene nanoribbons with edge cracks: a molecular dynamics study. Mol. Simul. **42**(14), 1157–1164 (2016). https://doi.org/10.1080/08927022.2016.1148266

62. Le, M.Q., Nguyen, D.T.: The role of defects in the tensile properties of silicene. Appl. Phys. A **118**(4), 1437–1445 (2015). https://doi.org/10.1007/s00339-014-8904-3

63. An, R.L., Wang, X.F., Vasilopoulos, P., Liu, Y.S., Chen, A.B., Dong, Y.J., Zhai, M.X.: Vacancy effects on electric and thermoelectric properties of zigzag silicene nanoribbons. J. Phys. Chem. C **118**(37), 21339–21346 (2014). https://doi.org/10.1021/jp506111a

64. Sahin, H., Sivek, J., Li, S., Partoens, B., Peeters, F.M.: Stone-Wales defects in silicene: formation, stability, and reactivity of defect sites. Phys. Rev. B **88**(4), 045434 (2013). https://doi.org/10.1103/PhysRevB.88.045434

65. Berdiyorov, G.R., Peeters, F.M.: Influence of vacancy defects on the thermal stability of silicene: a reactive molecular dynamics study. RSC Adv. **4**(3), 1133–1137 (2014). https://doi.org/10.1039/C3RA43487G

66. Noshin, M., Khan, A.I., Navid, I.A., Uddin, H.A., Subrina, S.: Impact of vacancies on the thermal conductivity of graphene nanoribbons: a molecular dynamics simulation study. AIP Adv. **7**(1), 015112 (2017). https://doi.org/10.1063/1.4974996

67. Gao, Y., Jing, Y., Liu, J., Li, X., Meng, Q.: Tunable thermal transport properties of graphene by single-vacancy point defect. Appl. Therm. Eng. **113**, 1419–1425 (2017). https://doi.org/10.1016/j.applthermaleng.2016.11.160

68. Yamamoto, K., Ishii, H., Kobayashi, N., Hirose, K.: Effects of vacancy defects on thermal conduction of silicon nanowire: nonequilibrium Green's function approach. Appl. Phys. Exp. **4**(8), 085001 (2011). https://doi.org/10.1143/APEX.4.085001

69. Dickey, J.M., Paskin, A.: Computer simulation of the lattice dynamics of solids. Phys. Rev. **188**(3), 1407 (1969). https://doi.org/10.1103/PhysRev.188.1407

70. Turney, J.E., McGaughey, A.J., Amon, C.H.: In-plane phonon transport in thin films. J. Appl. Phys. **107**(2), 024317 (2010). https://doi.org/10.1063/1.3296394

71. Fon, W., Schwab, K.C., Worlock, J.M., Roukes, M.L.: Phonon scattering mechanisms in suspended nanostructures from 4 to 40 K. Phys. Rev. B **66**(4), 045302 (2002). https://doi.org/10.1103/PhysRevB.66.045302

72. Hu, M., Zhang, X., Giapis, K.P., Poulikakos, D.: Thermal conductivity reduction in core-shell nanowires. Phys. Rev. B **84**(8), 085442 (2011). https://doi.org/10.1103/PhysRevB.84.085442

73. Delaire, O., May, A.F., McGuire, M.A., Porter, W.D., Lucas, M.S., Stone, M.B., Snyder, G.J.: Phonon density of states and heat capacity of $La_{3-x}Te_4$. Phys. Rev. B **80**(18), 184302 (2009). https://doi.org/10.1103/PhysRevB.80.184302

74. Yang, N., Zhang, G., Li, B.: Ultralow thermal conductivity of isotope-doped silicon nanowires. Nano Lett. **8**(1), 276–280 (2008). https://doi.org/10.1021/nl0725998

75. Li, N., Ren, J., Wang, L., Zhang, G., Hänggi, P., Li, B.: Phononics: manipulating heat flow with electronic analogs and beyond. Rev. Mod. Phys. **84**(3), 1045 (2012). https://doi.org/10.1103/RevModPhys.84.1045

76. Li, X., Chen, J., Yu, C., Zhang, G.: Comparison of isotope effects on thermal conductivity of graphene nanoribbons and carbon nanotubes. Appl. Phys. Lett. **103**(1), 013111 (2013). https://doi.org/10.1063/1.4813111

77. Chang, C.W., Fennimore, A.M., Afanasiev, A., Okawa, D., Ikuno, T., Garcia, H., Zettl, A.: Isotope effect on the thermal conductivity of boron nitride nanotubes. Phys. Rev. Lett. **97**(8), 085901 (2006). https://doi.org/10.1103/PhysRevLett.97.085901

78. Zhang, G., Li, B.: Thermal conductivity of nanotubes revisited: effects of chirality, isotope impurity, tube length, and temperature. J. Chem Phys. **123**(11), 114714 (2005). https://doi.org/10.1063/1.2036967

79. Cheng, Y.Q., Zhou, S.Y., Zhu, B.F.: Isotope effect on phonon spectra in single-walled carbon nanotubes. Phys. Rev. B **72**(3), 035410 (2005). https://doi.org/10.1103/PhysRevB.72.035410

80. Chen, S., Wu, Q., Mishra, C., Kang, J., Zhang, H., Cho, K., Ruoff, R.S.: Thermal conductivity of isotopically modified graphene. Nature Mater. **11**(3), 203 (2012). https://doi.org/10.1038/nmat3207

81. Jiang, J.W., Lan, J., Wang, J.S., Li, B.: Isotopic effects on the thermal conductivity of graphene nanoribbons: localization mechanism. J. Appl. Phys. **107**(5), 054314 (2010). https://doi.org/10.1063/1.3329541

82. Zhang, G., Zhang, Y.W.: Thermal conductivity of silicon nanowires: from fundamentals to phononic engineering. Phys. Status Solidi RRL **7**(10), 754–766 (2013). https://doi.org/10.1002/pssr.201307188

83. Jiang, J.W., Wang, J.S., Li, B.: A nonequilibrium Green's function study of thermoelectric properties in single-walled carbon nanotubes. J. Appl. Phys. **109**(1), 014326 (2011). https://doi.org/10.1063/1.3531573

84. Guo, Y., Zhou, S., Bai, Y., Zhao, J.: Tunable thermal conductivity of silicene by germanium doping. J. Supercond. Nov. Magn. **29**(3), 717–720 (2016). https://doi.org/10.1007/s10948-015-3305-1

85. Balasubramanian, G., Puri, I.K., Böhm, M.C., Leroy, F.: Thermal conductivity reduction through isotope substitution in nanomaterials: predictions from an analytical classical model and nonequilibrium molecular dynamics simulations. Nanoscale **3**(9), 3714–3720 (2011). https://doi.org/10.1039/C1NR10421G

86. Hu, J., Schiffli, S., Vallabhaneni, A., Ruan, X., Chen, Y.P.: Tuning the thermal conductivity of graphene nanoribbons by edge passivation and isotope engineering: a molecular dynamics study. Appl. Phys. Lett. **97**(13), 133107 (2010). https://doi.org/10.1063/1.3491267

Chapter 6
Summary and Concluding Remarks

Thermal transport in silicon-based nanomaterials has become a hot topic in the research community recently because it is one of the fundamental scientific problems in material physics and has enormous practical applications in modern electronics and thermoelectrics. Whereas thermoelectric materials which need the lowest possible thermal conductivity, thermal dissipation in micro/nanoelectronics requires the opposite. Over the past decades, rapid advances in fabricating silicon-based nanomaterials have attracted great interest among the scientific community to have a better understanding of thermal transport in these materials.

To promote silicon-based nanoscience and nanotechnology, we have performed systematic molecular dynamics simulations on thermal properties in low-dimensional silicon nanomaterials, including zero-dimensional nanoclusters and one-dimensional nanowires as well as two-dimensional crystal lattice with only a single atomic layer, such as silicene. Thermal transport properties in silicon nanostructures differ significantly from those in macrostructures because the characteristic length scales associated with heat carriers (i.e., the mean free path and the wavelength) are comparable to the characteristic length of nanostructures. The boundary, surfaces, and defects have been explored as routes to reduce thermal conductivity in silicon nanostructures, which is of particular interest for thermoelectrics applications.

Specifically, the thermal stability and thermal transport properties of silicon nanoclusters are strongly size dependent. For example, the melting temperature of silicon nanoclusters is lower than that of bulk silicon and increases with an increase in the cluster size. The structure changes upon heating, showing that the melting of silicon nanospheres is progressively developed from the surface into the core. The structure of the silicon nanocluster can change gradually from the bulk diamond structure to the amorphous structure with a decrease in the cluster size. Surface hydrogenation can significantly improve the energetic stability of the tetrahedral silicon nanoclusters, while silicon nanostructures without surface saturation are highly unstable and can

H.-P. Li and R.-Q. Zhang, *Phonon Thermal Transport in Silicon-Based Nanomaterials*, SpringerBriefs in Physics, https://doi.org/10.1007/978-981-13-2637-0_6

undergo amorphization in small-sized clusters. When the cluster size approaches the nanoscale, a significant reduction in the thermal conductivity can be induced from the stronger phonon-boundary scattering effects in smaller size, as well as from the stronger phonon-phonon scattering at higher temperatures. Regarding the thermal transport in silicon nanowires which is dominated by phonon-phonon coupling, numerous studies, including some simulations performed by us, highlight the importance of surface effects, such as surface roughness, surface functionalization, surface doping, surface disorder, and surface softening, for tunable thermal conductivity. The underlying mechanisms are attributed to the phonon-surface/boundary scattering, phonon-interface scattering, and phonon-defect scattering. It has been concluded that surface engineering methods are effective methods for modulating nanoscale thermal transport and may foster further advancements in the fields of silicon-based nanoelectronics and thermoelectrics. In addition, the remarkable success of graphene has stimulated renewed interest in graphene-like 2D layered nanomaterials. Among them, silicene—the silicon-based counterpart of graphene—has a 2D buckled structure that is responsible for a variety of potentially useful physical properties. Following our pioneer simulation of the phonon thermal transport of pristine silicene sheets, many strategies have been studied in the research community to reduce thermal transport in silicene and nanoribbions, including adding substrates, heterostructuring, surface functionalization, and defects, providing significant guidance for experimental realization. Our recent investigations also show that the effects of vacancy defects and isotope doping induce the significant diminution in the thermal conductivity of silicene sheets. Particularly, phonon thermal transport in silicene is strongly affected by the vacancy concentration, vacancy size, and vacancy boundary shape. Also, the phonon thermal conductivity of isotopic doped silicene nanoribbons is dependent on the concentration and arrangement pattern of dopants. The findings presented in this book could be helpful in furthering the understanding of phonon thermal transport in silicon-based nanomaterials and may serve as a highly useful experimental guide in silicon-based thermoelectric applications as well as in other thermal-related applications.

Nonetheless, there are still challenges and debates—both experimentally and theoretically—that deserve further investigation in this field. For example, although the thermal conductivities of silicon nanoclusters and silicene sheets have been predicted in theory, there is a lack of accurate experimental measurements. The precise mechanisms that underlie the surface-induced reduction of thermal conductivity remain unknown. In particular, the synergistic effects of the material surface and other factors on the thermal transport in silicon nanowires remain unclear. Measuring thermal conductivity at the nanoscale continues to face many difficulties, such as determining thermal contact resistance between the sample and substrate, and also differs from group to group, and there is heated debate over the effects resulting from sizes, defects, and surfaces. In addition, three levels of theoretical and computational approaches—microscopic level, mesoscopic level, and macroscopic level—have been used to describe nanoscale phonon transport, but at present, there are limitations to the application range of these different level methods. Therefore, in parallel with the experimental research, it is essential and urgent to develop new theories and methods to rapidly and accurately calculate the thermal conductivity (including phonon and electron contributions) of nanomaterials in the recent near future.

Index

B
Ballistic phonon transport, 13

D
Defects vacancy, 70, 72
Diffusive phonon transport, 13

E
Electron-beam heating technique, 29, 30

G
Green-Kubo method, 20–22

H
Heat current autocorrelation functions, 46
Hydrodynamics, 16, 18
Hydrogenation, 45, 81

I
Isotope doping, 73, 82

M
Melting, 41–44, 81
Molecular dynamics simulations, 19, 81
Müller-Plathe method, 24, 25

N
Non-equilibrium Green's function method (NEGF), 26
Normal scattering, 13

O
Optothermal Raman technique, 27, 28, 30, 31
Ordered doping, 75

P
Phonon, 2, 5, 6, 12–20, 26, 27, 33, 34, 46–49, 54–61, 69–76, 82, 83
Phonon Boltzmann transport equation, 16, 17, 69
Phonon density of states, 2
Phonon mean free path, 13, 14, 48, 54

R
Random doping, 75
Resistive scattering, 13

S
Silicene, 6, 34, 67–73, 81, 82
Silicene nanoribbons (SNRs), 67, 69, 73, 82
Silicon nanoclusters, 6, 42, 44–48, 81, 82
Silicon nanowires (SiNWs), 5, 6, 27, 31, 33, 53, 70, 74, 82
Simulated annealing, 44–46
Size dependence, 43, 44, 48
Structural transition, 44
Surface disorder, 55, 59, 82
Surface doping, 82
Surface effects, 6, 54, 55, 82
Surface functionalization, 6, 53, 55, 57, 70, 82
Surface roughness, 5, 27, 53–56, 59, 82
Surface softening, 55, 82

Printed in the United States
By Bookmasters